Thank you for returning
your books on time.

To a Distant Day

Outward Odyssey
A People's History of Spaceflight

Series editor: Colin Burgess

TO A DISTANT DAY

The Rocket Pioneers

CHRIS GAINOR | FOREWORD BY ALFRED WORDEN

UNIVERSITY OF NEBRASKA PRESS • LINCOLN AND LONDON

© 2008 by the Board of Regents of the University of Nebraska ¶ All rights reserved ¶ Manufactured in the United States of America ¶ ∞ ¶ Library of Congress Cataloging-in-Publication Data ¶ Gainor, Chris. ¶ To a distant day : the rocket pioneers / Chris Gainor; foreword by Alfred Worden. ¶ p. cm. — ¶ (Outward odyssey : a people's history of spaceflight) ¶ Includes bibliographical references and index. ¶ ISBN 978-0-8032-2209-0 (cloth : alk. paper) ¶ 1. Rocketry— History ¶ 2. Astronautics—History. ¶ I. Title. ¶ TL781. G25 2008 ¶ 621.43'560922—dc22 ¶ 2007029365 ¶ Set in Adobe Garamond and Futura by Kim Essman. ¶ Designed by R. W. Boeche. ¶

To the men and women whose work behind the scenes
makes possible the flights of humans into space.

Contents

Illustrations

Foreword

Alfred Worden
Apollo 15 Astronaut

I was born and raised, along with five brothers and sisters, on a small farm in Michigan. By the time I was twelve I was working full days in the fields and milking cows. This was just after the Great Depression and at a time before space travel, before jet engines, in a year when the world's most sophisticated rockets were generally not getting much higher than a couple of hundred feet—if they didn't blow up first.

It would have been hard for me to imagine that only two decades after leaving that farm, I'd be heading into space on a rocket as large as a forty-story building—a rocket so huge that I was a long, long way above the ground even *before* we lifted off. It was so tall, in fact, that it was not a noisy ride—because the noise was generated about three hundred feet below us, about the length of a football field. But I could feel the force of rocketing into orbit. As the fuel burned out of the first stage and the vehicle became lighter we accelerated to about 4 g's, which was like four people sitting on me, and it was not pleasant. The Saturn V rocket, which took three of us all the way to the moon, was an incredibly accurate machine, operating to the precise split second planned. Our crew spent a lot of time training in case something went wrong during launch: for example, if the engines quit, an escape tower would pull us away to safety. But we never needed that training—the rocket worked just fine. I even wrote a couple of poems about that beautiful, majestic Saturn V rocket.

The *Apollo 15* mission that I flew has been called the high point of human exploration. The rocket launch was just the beginning of the lunar journey, but I'd already taken a long journey in my career to get to that point.

As a teenager, working on that farm, I realized that farming was not what I wanted to do for the rest of my life. I decided that I wanted to go to college, and I ended up with the equivalent of an engineering degree from the United States Military Academy at West Point. I still didn't have any real interest in flying or rocketry, but decided to go into the air force because I heard that the promotions were quicker there than in the army. I was wrong about that, but in the process I found out something else. I got really interested in how to fly an airplane, especially flying on instruments. I soon realized that I had a gift for it; I could do it pretty well. Flying airplanes is not the same as flying rockets, but it wasn't long until rocketry also became part of my career.

My first real interaction with rockets was when I was assigned to an air defense squadron in Washington DC. We were flying live missiles and rockets on those aircraft in defense of the nation's capital, and without a lot of flying time available I got really interested in the maintenance activity going on. Soon, I was running the section: we rebuilt the armament shop and turned it into a modern repair shop like you see today. Air Defense Command liked it so much they wanted me to head up a program to help other fighter squadrons do the same thing. Instead, I went back to college, this time to learn about guided missiles. That led me into being a test pilot—the top of the heap in aviation.

I ended up teaching as well as flying, at the test pilot school at Edwards Air Force Base. For a while I was training the pilots on the X-15 rocket plane in landing techniques. I was even training astronauts like Gene Cernan and Charlie Bassett in a zero-gravity simulator that we had. I wrote and taught a number of the courses in spaceflight that the test pilots had to pass; my transition from flying planes to flying rockets was getting ever closer.

Although it was always something to work toward, I hadn't thought about being an astronaut; when I became a test pilot, there just weren't any openings. But after about a year and a half, they had a selection. I'd always thought of it like winning the lottery—it would be nice, but was never going to happen. Too many other people wanted it. We even fooled someone else in our squadron into thinking he was selected. We kept that prank going for about two weeks until we told the poor guy. I was very fortunate—with the right combination of college work and test pilot experience, I made the cut.

I soon realized that being a test pilot, the top of my profession, counted for nothing in the astronaut circle. I was one of the new, green kids on the block. I understood the game—I had to get on a rocket and make a flight before anyone would consider me a "real" astronaut. It was pretty humbling. Individually, I knew I had a lot of very hard work ahead of me, and collectively our new astronaut group really applied itself to the mammoth task ahead. During the first week, while undergoing instruction in orbital mechanics and rocket trajectories, we came to the realization that we could probably do a better job training ourselves. Our recommendation was accepted and a period of successful self-instruction began. Soon, I was working on the Apollo Command Modules as they were being built, and got to know their workings probably better than anybody.

I was on the support crew for *Apollo 9*, on the backup crew for the *Apollo 12* moon landing mission, and then flew on *Apollo 15*. It was a lot of work, but a lot of fun too. We were launching so frequently there was not a lot of time between missions. Yet by the end of the training we felt that we could handle any malfunctions, any problems.

Our *Apollo 15* mission truly *was* exploration at its greatest. Unlike previous lunar missions, where they only had limited time, we had better spacesuits, an upgraded Lunar Module that allowed a longer stay, and even a lunar rover, allowing my two crewmates to drive around on the surface. This lengthier mission—twelve days in space—also meant that I was in lunar orbit longer than anyone had ever been before. After that Saturn rocket took us to lunar orbit I spent three days alone, a quarter of a million miles from earth, mapping the surface of the moon and operating remote sensors from the orbiting Command and Service Module.

I would have liked to have made a lunar landing, no question about that. But I was very happy where I was. In my thirty-five solo orbits of the moon, I was able to thoroughly observe a quarter of its surface, compared to the small area my crewmates were able to traverse after landing. It was a long way from home, and the only thing connecting me to earth was the radio. It was even a little scary. But after four days in a crowded spacecraft with two other guys, it was certainly nice to be on my own. If it wasn't for enormous and precisely functioning rockets, we would never have been able to achieve this incredible scientific exploration. But I was always aware of the risks we were taking. Before the launch, I talked to my family, and I remember thinking that this might be my goodbye: If the rocket malfunctioned, I might not be coming back to see them again. We astronauts accepted the risk because the rewards were worth it. My only real concern was if something did go wrong, I didn't want to be the one who had caused it.

Having flown on what many consider the most successful scientific flight of the entire Apollo program, I have a rather unique perspective on the history of rocketry. Riding that Saturn V, I was particularly appreciative of all those pioneers who developed rockets over the centuries. They came from a huge variety of backgrounds and nations, and most knew failure more than they ever enjoyed success. But each, in his own way, pushed what was possible into new realms of discovery. Their stories are unique and fascinating, and I hope that you enjoy reading them in this absorbing book. It's just the beginning of a long adventure, and a prelude to the day, hopefully not too long from now, when we push ever deeper into space in humankind's outward odyssey.

Preface and Acknowledgments

This book got its start in 2003 when Colin Burgess, editor for the Outward Odyssey series of books on the history of space exploration, contacted me to ask if I was interested in writing a history of space exploration leading up to the time of the first human spaceflight in 1961. I had never thought of writing such a book, but as I considered the proposition and looked into the available literature, I warmed to the idea. One big reason was that our understanding of the events covered in this book has changed significantly in the past few years. The fall of the Soviet Union and the opening of its archives were factors in these changes, but as I looked around, I realized that historians were also challenging the established views of the early space pioneers from other parts of the world.

While writing this book, this recent scholarship caused me to reconsider my own views and understanding of many twentieth-century space efforts. I hope that readers of this book will gain a new appreciation of the events and individuals that led to the space race of the 1960s, and of the events that followed, which will be covered in other books in the Outward Odyssey series.

I also accepted Colin's commission because I had spent much of the past decade engaged in studies of the history of space exploration, first for a book I wrote on the Canadian and British engineers who joined the U.S. space program in 1959 when the Canadian government canceled a military aircraft program. When that book was complete, I began to pursue academic studies. During that time, I have been pondering a question that has as much to do with the future of space exploration as its past. In the 1960s, Apollo's journeys to the moon seemed to be a natural part of human progress. Many other people took it for granted that humans would be colonizing the moon and walking on Mars by the end of the twentieth century. As we now know, things didn't turn out that way. This turn of events inspires questions about what really led to the space race of the 1960s. I invite readers to consider how the events and social forces discussed in this book have affected developments in space exploration in the 1960s and since.

A key part of the story in this book is the wave of interest in space exploration that swept Europe and particularly Germany in the 1920s and 1930s. Recent studies by Frank H. Winter and David Clary suggest that Robert Goddard's biggest contribution to space exploration was not necessarily his creation of the first liquid-fuel rocket, but instead his being the first credible scientific figure to point to the rocket as the way into space. Goddard was soon followed by Hermann Oberth and Max Valier, who spread the word in Germany, and by many Russians, especially Friedrich Tsander, who publicized Konstantin Tsiolkovsky's contributions to spaceflight theory. The Frenchman Robert Esnault-Pelterie, who in my view stands together with Goddard, Oberth, and Tsiolkovsky as an important early theorist of spaceflight, also helped fan the early flames of enthusiasm.

As Goddard learned the hard way, progress in rocketry requires large teams of experts, not solitary inventors. In the 1930s, the German army assembled the first team of engineers and scientists dedicated to building large rockets. Even though their ultimate product, the V-2 rocket, proved to be of extremely limited value as a weapon, the German rocketeers continued to receive support from the government of Nazi Germany. Had the German army flagged in building the V-2, the subsequent history of space exploration would be very different. But it didn't, and World War II ended with the creation of the V-2 by Germany and the atomic bomb by the United States. Soon the Soviet Union had its own nuclear weapons and began building rockets as a means to deliver these weapons to its Cold War adversary. The United States also developed its own rockets capable of delivering nuclear weapons. Soon both sides possessed rockets that were also useful for launching vehicles into orbit around the earth and into deep space. Barely four decades after rockets were acknowledged as the way into space, humans began using them for that purpose.

While I did my research I came across many individuals who made huge contributions to space travel but received little recognition. Esnault-Pelterie comes to mind, as do Valier and Tsander. In this book, I have mentioned these individuals and many other people who contributed enormously to spaceflight in Germany, Russia, America, and even in the smaller space powers that began to emerge in the 1950s.

I used a large number of sources, which are listed at the end of this book, and I invite readers seeking more information and insight to consult them. Many "prehistories" of space exploration were written in the 1960s and 1970s, when the space race between the United States and the Soviet Union was at its height. Books by Willy Ley, the great German-American popularizer of spaceflight who died just weeks before *Apollo 11*, immediately come to mind. His *Rockets, Missiles, and Men in Space* turned out to be a valuable resource for me while writing this book. So was Wernher von Braun and Frederick I. Ordway's *History of Rocketry and Space Travel*, and a volume edited by U.S. historian Eugene Emme, *The History of Rocket Technology*. Many other books about the pioneers of space travel, such as Konstantin Tsiolkovsky, Hermann Oberth, and Robert Goddard, also date from this time. David Clary's recent biography, *Rocket Man*, provides important new perspectives on Goddard's work.

I have mentioned the fall of the Soviet Union in 1991. One of the many consequences of that event was the opening of Soviet archives and the freeing of Soviet participants in space and missile programs to speak their minds. Among the historians parting the curtains on the history of Soviet space efforts is Asif Siddiqi, whose *Challenge to Apollo: The Soviet Union and the Space Race, 1945–1974*, has become the indispensable reference on its subject. Siddiqi has also written many articles and is now producing an English-language version of Soviet space pioneer Boris Chertok's memoirs. Siddiqi's work proved invaluable for me, and many other histories from inside and outside Russia were also great resources.

A great American institution headquartered in Washington DC played an important role in the events covered in this book. The Smithsonian Institution supported Robert Goddard's early rocket work, and today its National Air and Space Museum (NASM) continues to support historians who write important histories of space exploration. NASM historian Frank H. Winter has written the definitive account of the rocket societies of the interwar years, *Prelude to the Space Age: The Rocket Societies: 1924–1940*. While I was writing this book, I met him at the 2004 International Astronautical Congress in Vancouver, and he shared with me a paper he is writing on Goddard that contains many important insights on Goddard's contribution to space exploration. NASM historian Michael J. Neufeld's *The Rocket and the Reich* is the best history available on the German rocket program of the 1930s and 1940s. I have exploited the rich vein of literature on von Braun and his rocket team. Since much of it is dated or takes a strong stand one way or another on the involvement of von Braun and his colleagues in the Nazi atrocities of World War II, Neufeld's thorough and independent scholarship is most helpful.

I also made use of titles produced for the NASA history series. NASA's support of history speaks highly of the importance that institution places on disseminating the findings of its pioneering efforts.

For the history of the Jet Propulsion Laboratory, I relied on Clayton R. Koppes's *JPL and the American Space Program*, and Iris Chang's memorable biography of Tsien Hsue-shen, *Thread of the Silkworm*. Paul Dickson's *Sputnik* expertly covers the reaction that historic satellite instigated. Craig Ryan's *The Pre-Astronauts* chronicles the balloon flights of the 1950s.

I also referred often to two recently written general histories of space exploration, William E. Burrow's *This New Ocean* and T. A. Heppenheimer's *Countdown: A History of Space Flight*, along with many works by Roger Launius, who is now at the NASM. A vital source of general guidance was Walter A. McDougall's groundbreaking political history, . . . *The Heavens and The Earth*. William Sims Bainbridge's provocative study *The Spaceflight Revolution* also gave me much food for thought while I was writing this book.

Many individuals have helped me with this study. Colin Burgess selected me to write it and provided advice and encouragement throughout the writing process, including a close reading of the completed manuscript. He also shared with me a monograph he wrote in anticipation of a book on animal flights into space. I would also like to thank *Apollo 15* astronaut Col. Alfred Worden for his wonderful foreword.

I began on this book just after completing my master's degree in the Space Studies Program at the University of North Dakota. My professors there, Shanaka de Silva, Mike Gaffey, Eligar Sadeh, John F. Graham, and especially Stephen B. Johnson, provided countless insights into their subjects that inform my writing. I am now engaged in further studies at the University of Alberta, and I would like to thank my faculty advisor, Dr. Robert W. Smith, for his guidance and assistance. My studies in research methods with Dr. Guy Thompson were also helpful.

I received assistance from many friends, including Bruce Baugh, Rolf Maurer, Steve Pacholuk, Ken Harman, and Barry Shanko. While I was putting the finishing touches on this book, I received the news that my friend and space enthusiast, William "Butch" Head of Beech Grove, Indiana, had passed away. I had looked forward to sharing this book with Butch.

Finally, my family. My brother, Mark Gainor, and sister, Mary Gainor, provided invaluable help to me in pursuing my academic studies and this book. My brother, Tim Gainor, also helped. Keyano kept me company. My parents, Don and Toni Gainor, as always, gave me encouragement. None of my work would be possible without the support and understanding of my wife, Audrey McClellan.

Acronyms and Abbreviations

A-1— Aggregat (or Assembly), German series of rockets leading to the A-4 and later models

ABMA— Army Ballistic Missile Agency

ARPA— Advanced Research Programs Agency; today it is known as DARPA, the Defense Advanced Research Programs Agency

ARS— American Rocket Society

BIS— British Interplanetary Society

Caltech— California Institute of Technology

COPUOS— United Nations Ad Hoc Committee on the Peaceful Uses of Outer Space

COSPAR— Committee on Space Research

FAI— Fédération Aéronautique Internationale

FBI— Federal Bureau of Investigation

GALCIT— Guggenheim Aeronautical Laboratory, California Institute of Technology

GDL— Gas Dynamics Laboratory, Leningrad

GIRD— Russian acronym for the Group for the Study of Reactive Propulsion

GPS— Global Positioning System

IBM— International Business Machines

ICBM— intercontinental ballistic missile

IGY— International Geophysical Year

IRBM— intermediate-range ballistic missile

JATO— jet (actually rocket)-assisted take-off

JPL— Jet Propulsion Laboratory, Pasadena, California

LOR— lunar orbit rendezvous

MIT— Massachusetts Institute of Technology

MOUSE— Minimum Orbital Unmanned Satellite, Earth, a 1953 satellite proposal

MSC— Manned Spacecraft Center, Houston, Texas (1961 to 1973)

NACA— National Advisory Committee for Aeronautics

NASA— National Aeronautics and Space Administration

NERVA— nuclear engine for rocket vehicle applications

NII— Russian acronym for Scientific Research Institute

NKVD— Russian acronym for the People's Commissariat for Internal Affairs, the Soviet secret police of the Stalin era, later known as the KGB

NOTS— Naval Ordnance Test Station, China Lake, California

OKB— Russian acronym for Special Design Bureau; Sergei Korolev headed OKB-1

PARD— Pilotless Aircraft Research Division

PSAC— President's Science Advisory Committee

R-1— First of Soviet family of rockets leading to the R-7 ICBM and beyond

RAF— Royal Air Force

RAND Corporation—Originated as Project RAND (for Research And Development) at Douglas Aircraft, but RAND split off to become an independent "think tank"

REP— Robert Esnault-Pelterie

RNII— Russian acronym for Reactive or Rocket Scientific Research Institute

SCORE— Signal Communications by Orbiting Relay Equipment

SS— Schutzstaffel, the notorious Nazi paramilitary organization

STG— Space Task Group, forerunner to the Johnson Space Center

TCP— Technological Capabilities Panel

TIROS— Television Infrared Observation Satellite, the first weather satellite

TRW — Thompson Ramo Wooldridge Inc.

U-2 — High altitude U.S. surveillance aircraft

USAF — United States Air Force

USSR — Union of Soviet Socialist Republics

V-1, V-2 — Vergeltungswaffe (German revenge weapon of World War II)

VfR — Verein für Raumschiffahrt (German Rocket Society)

WAC Corporal — The "little sister" rocket of the Corporal, WAC stood for Women's Army Corps (or "without attitude control")

To a Distant Day

1. Space Dreams and War Drums

The earth is the cradle of mankind, but one cannot stay in the cradle forever.

Russian spaceflight theorist Konstantin Tsiolkovsky

This famous assertion by Tsiolkovsky is always cited to support the idea that humanity must go into space to survive and flourish. Yet few people have noted the implications of the first part of his statement. Instead of saying humans originated on the earth, he called the earth our "cradle." Tsiolkovsky chose his words to reflect his belief that life is widespread throughout the universe and that earthly life didn't necessarily start on the earth. And the explosion of scientific knowledge in the twentieth century suggests that in this respect Tsiolkovsky might be correct. Consider this fact: The chemical composition of humans and other living things more closely resembles the stars than the earth. The four most common elements in living beings are hydrogen, oxygen, carbon, and nitrogen. Scientists have determined that these four elements are the four most common elements in the universe, except for helium and neon, which don't form compounds. The four most common elements on earth are silicon, iron, magnesium, and oxygen.

Some researchers have gone so far as to suggest that comets, meteorites, or other debris from space "seeded" life on earth, but this idea remains controversial at best. Yet it is clear that forces that came from beyond earth's atmosphere affected the long and difficult evolution of earthly life—humanity's time in the cradle. The moon's impact on earthly life through tides and biological cycles is well known. In recent years, scientists have tied the mass extinction of the dinosaurs sixty-five million years ago to the impact of a gigantic asteroid that caused major climate change on earth. In the more than three billion years that life has existed on earth, there have been more than a dozen mass extinctions, and many of them may be related to similar impacts by comets or asteroids. Other researchers are looking into aspects of radiation and other "weather" in space that could have affected life on earth.

Humankind's reach into space is changing our view of the universe and our place in it. It remains to be seen whether it will be necessary to establish colonies beyond earth to save humanity and other earthly life from some future catastrophe. The answers to such questions are beyond the scope of this book, but the fact that they are being seriously raised represents just the latest thrust in a long evolution of human thought that at first barely acknowledged the existence of anything beyond earth.

We can only conjecture what our early ancestors thought they saw when they looked up. There are many suggestions that they saw the sky simply as a continuation or reflection of what they saw on the ground. The design of the great megalithic monuments such as Stonehenge strongly imply that their builders had a good knowledge of the movements of the sun and moon. Astronomy and astrology were in evidence in the earliest civilizations, which began between five thousand and six thousand years ago. The Babylonians who lived in today's Iraq studied the skies as they did every aspect of nature in their search for omens that might warn of a disaster that could be avoided by performing an appropriate ritual. The motions of the moon, the sun, and planets were recorded by Babylonian astronomers not only to use as omens but in developing calendars and to predict celestial phenomena. Ancient Egyptians observed the sky to develop their own calendars. Most importantly, they used the appearance of the star Sirius in the sky to predict when the Nile would flood, the annual event around which Egyptian life revolved.

Among their many gifts to civilization, the Greeks advanced astronomy. The mathematician Pythagoras and his band of mystic followers in the sixth century B.C. believed in a universe based on mathematics. They correctly viewed the earth as a sphere, and Eratosthenes, who lived in Egypt in the third century B.C., made a reasonably accurate estimation of the earth's size. Aristarchus and others believed that the earth orbited the sun, but in the end, Greek science put the earth at the center of a universe bounded by a larger sphere to which the stars were attached. They believed that the sun, moon, and the wandering stars known as planets were attached to their own spheres that rotated about the earth. These beliefs, as propounded by the philosopher Aristotle and codified by the second-century B.C. astronomer Ptolemy, persisted for more than fourteen hundred years with the support of the Catholic Church, even though Ptolemy did not consider his work to be the final word on the subject. Though Aristotle believed firmly that there were no worlds beyond the earth, Plutarch, who lived in the first and second centuries A.D., wrote of the moon as a world unto itself.

The Greeks are generally credited as being the first to think of humans taking flight, and the first to suggest that there are other worlds to visit. The myth of Daedalus and his son Icarus involved a winged escape from a blockaded Crete. But Icarus died when he flew too close to the sun, and the wax holding his wings together melted. The great philosopher Socrates was quoted as saying: "Man must rise above the earth—to the top of the atmosphere and beyond—for only thus will he understand the world in which he lives." Lucian of Samosata, a second-century A.D. satirist, wrote two stories that involved trips to the moon. The first, "A True Story," involved a voyage on a sailing ship that left the Mediterranean Sea for the Atlantic Ocean, and was intended to poke fun at the tales of the terrors of the Atlantic that Phoenicians told to discourage others from sailing there. At one point, his ship is lifted by a whirlwind, and after being airborne for seven days and seven nights, it alights in a cultivated land inhabited by men riding three-headed vultures who tell them they are on the moon. Like the real astronauts of the twentieth century, Lucian's space voyagers looked back. "We also saw another country below, with cities in it and rivers and seas and forests and mountains. This we inferred to be our own world." Lucian's voyagers were soon caught up in a war between the inhabitants of the sun and the moon in which both armies were made up of men and various fantastic animals. Lucian's men are taken prisoners by the forces of the sun, and the moon surrenders after the sun men build a wall that puts the moon in darkness. Lucian imagines that stars and comets are inhabited, but finds no women on any of these bodies. He recounts that in the end his voyagers are allowed to return to earth, where they go on to further adventures.

Lucian's other story, "Icaromenippus, or the Sky-Man," involved the third-century B.C. philosopher Menippus, who in this story decides to take flight using an eagle's wing and a vulture's wing. Since there is no wax involved, he avoids Icarus's fate, and flies to the moon. Menippus looks back at the earth, but almost fails to recognize it. Flying on to heaven, he meets with the gods and has adventures that satirize the views of the philosophers, until Zeus orders Menippus's wings confiscated, and he returns to earth. Though it is possible that some of Lucian's stories were based on earlier writings, including those of Menippus himself, his stories marked the first appearance of a literary theme that has persisted to this day: the use of space as an empty canvas useful for painting fantastic tales.

Astronomers and navigators in the Arab world, China, and India continued looking to the sky and recording what they saw, while science in Europe stagnated until the invention of printing in the fifteenth century stimulated the arts and sciences. Throughout this period, the planets were seen as starlike objects whose forward and backward motions among the fixed background of the stars required explanation. Most educated people relied on the complicated Greek concepts of epicycles and deferents to explain and predict planetary motion on the assumption that the earth lay at the center of the universe. But the Polish canon and astronomer Nicholaus Copernicus challenged this view in his groundbreaking work, *De Revolutionibus Orbium Coelesticum* (On the Revolutions of the Heavenly Spheres), which was published as Copernicus lay on his deathbed in 1543. In this work, Copernicus showed that the earth and the other planets orbited around the sun, and that the moon alone orbited the earth.

Copernicus's work was not widely accepted for many years after his death, but during the sixteenth century the work of Tycho Brahe and Johannes Kepler continued to advance astronomical knowledge. Tycho's observations of the comet of 1577, for example, showed that comets were not atmospheric phenomena but objects in outer space. Tycho also noted that the 1577 comet had moved within the planetary regions, disproving the idea that planets were attached to spheres. Based on Tycho's accurate observations of planetary motions, which questioned Ptolemaic and Copernican predictions of planetary motion, Kepler wrote that planets and other bodies in space move along elliptical and not circular paths.

In 1609, a forty-five-year-old Italian mathematics professor named Galileo Galilei got word of the invention in Holland of a new device that could magnify distant objects. Galileo began building his own versions of this device, which became known as a telescope, and he became the first person to point it at the sky and publish his findings. On January 7, 1610, Galileo saw that Jupiter was accompanied by three small points of light. On succeeding nights, he found that there were four small "stars" that stuck with Jupiter in a way that could only be explained if they were moons of Jupiter. He also pointed his telescope at our moon, a body that was considered perfect in spite of the irregular pattern on its face, which was thought to be a reflection of the earth. Instead of a perfect sphere, Galileo found mountains and other irregular structures on the lunar surface. When he looked at Saturn, he saw two strange appendages that later observers determined to be its rings, and he saw that Venus had phases like the moon. These findings led Galileo to champion the Copernican view of the universe and brought him in conflict with the Catholic Church. But most importantly, Galileo and the other observers of the seventeenth century who eagerly followed him established that there were distinct worlds beyond the earth.

The finding that there were indeed other worlds stimulated people's imaginations, and many of those imaginings soon found their way into print. Kepler, for his part, wrote a work of fantasy that was published after his death in 1630, the *Somnium*, whose descriptions of the moon benefited from telescopic observations of the lunar surface. At roughly the same time, two English bishops wrote books about the moon. Bishop Francis Godwin's fictitious hero used swans to take him to a moon inhabited by large humans who communicated using a musical language. Bishop John Wilkins wrote a more serious book suggesting that the moon was similar to the earth, could be inhabited, and could be visited by humans. In the middle of the seventeenth century, Cyrano de Bergerac wrote two books about space travel. The first one, *Voyage dans la Lune*, depicted a voyage to the moon, and the second, *Histoire des ftats et Empires du Soleil*, took its readers to the sun. In his stories, De Bergerac tried a number of devices to get into space, including bottles filled with dew that would be carried away by the rays of the morning sun, and finally a flying machine with rockets attached to it. Though these were part of a long string of fictional accounts of trips to the moon, the sun, and the planets, de Bergerac was the first to hit on the method that would actually open the way for space three hundred years later.

The Rockets' Red Glare

By the time of de Bergerac's books, rockets were long established as fireworks and more importantly, as weapons, even though the principles behind their operation weren't yet fully understood. Most historians agree that rockets originated with the invention of black powder, known since the invention of guns as gunpowder. This powder, a mixture of charcoal, sulfur, and saltpeter, is believed to have been invented by alchemists in China between fourteen hundred and two thousand years ago, and was soon adopted for fireworks and then for explosives. Fixing an approximate time for the invention of rockets is more difficult, because historical accounts of battles speak of weapons that could be either rockets or simply flaming arrows whose heads were dipped in a flammable substance and set alight before firing. It is clear that by the middle of the eleventh century, the Chinese were well acquainted with the properties of black powder, and its use in rockets likely dates back to this time.

Mongol armies are known to have used gunpowder in Europe in 1241 and rockets in an attack on Baghdad in 1258. Around this time, the Arabs also began to employ rockets as weapons, and the use of black powder quickly spread to Europe. The use of rockets in India is traced back to 1399. As the years passed, more people studied gunpowder and rockets in an effort to make them more effective as weapons and for fireworks displays. The Italian historian Muratori first used the word *rocket* in its Italian form *rochetta*, referring to its cylindrical shape, in 1379, and Italians advanced the art of fireworks in the years that followed. A well-known legend, which no evidence survives to substantiate, has it that the first human attempt to ride a rocket took place about 1500 in China. Wan Hu, a Chinese mandarin, is said to have disappeared in a cloud of smoke after assistants lit forty-seven rockets attached to his chair.

Rockets were used from time to time in battle in Europe and elsewhere after the thirteenth century, though not systematically because of their limited range and accuracy. In the 1790s, Hyder Ali, prince of Mysore, and his son Tipu Sahib, used rockets effectively against British forces in India. These rockets were made of an iron tube instead of the then typical pasteboard, and their range was about one thousand yards. Reports of these rockets impressed an English socialite and inventor who had never been to India: William Congreve.

Starting about 1801 with some fireworks rockets he purchased in London, Congreve, a former lawyer and newspaperman who had studied at Cambridge, began his work to make better rockets at the Royal Laboratory at

Woolwich, where his father held a high position. The British military, concerned by the threat of an invasion from France by Napoleon's forces, soon started to support Congreve's research. Congreve not only created a more advanced rocket; he developed what Smithsonian historian Frank H. Winter has termed a "complete weapons system." By refining the gunpowder and using special presses to pack it tighter into metal cases, Congreve obtained longer distances and better accuracy from his rockets. Congreve provided a whole family of rockets of varying calibers, ranges, and sizes that could be fitted with warheads and used to explode, spread shrapnel, set fires, or illuminate darkened battlefields. The rockets came with launching stands, and Congreve even supervised the fitting of special ships from which rockets could be launched.

Congreve's rockets were soon put to use, first at Boulogne, where Napoleon's fleet for the invasion of England was located. The first British naval attack with rockets on this French port was called off in 1805 due to poor weather, but on a second try the following year, the rockets put the city on fire. Similar rocket attacks in Copenhagen in 1807 and against Napoleon's forces at the Battle of the Nations in Leipzig in 1813 were also effective. British forces successfully used Congreve's rockets in North America in the War of 1812. Their most memorable moment came in circumstances in which the rockets did not prove decisive for the British. During an attack in September 1814 on Fort McHenry at Baltimore, the sight of "the rockets' red glare" illuminating an American flag inspired a lawyer named Francis Scott Key to write a poem called "The Star Spangled Banner," which was later set to music and became the national anthem of the United States. The fort held against the rockets and artillery of the British in what was a turning point in the war for the Americans.

All rockets are reactive devices, which means that they generate an action that causes an equal and opposite reaction, in this case the movement of the rocket. This action-reaction principle is the third law of motion spelled out by Isaac Newton in his 1687 masterwork, the *Principia*. Building on the work of Kepler and others, Newton explained gravity and the motion of bodies in the universe. Though most people believed that rockets worked by pushing against air, a belief that persisted into the twentieth century, Newton's laws showed that rockets could work anywhere, with or without air. For the military in the nineteenth century, this meant that rockets didn't recoil like cannons or other guns did, which is why Congreve equipped ships with rockets to attack port cities. No matter how many rockets were fired, the ships didn't rock as they did when cannons were fired from their decks.

In spite of Congreve's improvements, rockets still had their limitations and were best for specific situations. For example, rockets that could set towns ablaze were not useful in cases where an army wished to win over a local population. They were also more effective against unsophisticated enemies who weren't prepared for rockets or didn't know what they were. And there were still limitations on their range and accuracy. Advances in guns and artillery also caused rockets to go out of fashion until rockets themselves were improved.

Congreve, who arguably was the world's first rocket scientist, worked on perfecting rockets until his death in 1828. Though he was inspired to advance rocketry for military purposes, he also thought of flying beyond the earth. As a child, he wrote of his wish to fly to the moon in the recently invented hot-air balloon, and late in life he wrote of an "Aerial Carriage" powered by wind and muscle.

Rockets came to be used by most European armies and navies at various times in the nineteenth century. The Russian military began developing the weapons in 1810 and used them effectively in the Russo-Turkish War of 1828–29 as well as in colonial wars, including engagements in 1853 in Central Asia not far from the place that became Russia's major spaceport a century later. In 1834, Karl A. Shilder developed a submarine for the Russians that fired rockets, presaging a major weapon of the nuclear age 130 years later. In the Crimean War of 1853–56, both the Russians and their English, French, and German opponents fired rockets at each other. The U.S. Army used rockets in the Mexican War of 1846–48 and in the Civil War. A demonstration of rocketry for President Abraham Lincoln in 1862 almost turned the tide of history when a rocket exploded near the president, though luckily it didn't harm him.

For all their advances, Congreve's rockets still carried long sticks to direct them at launch. Another British inventor, William Hale, found in the middle of the century that rockets could be stabilized by spinning them during flight. Three metal vanes inserted in the exhaust nozzle and tilted slightly toward the direction of the exhaust caused the rockets to rotate. Hale also built other improvements into rockets, including a hydraulic press for the gunpowder fuel. As the nineteenth century wore on, rockets were brought into use to harpoon whales, to signal distress, and to throw rescue lines. Their use in battle declined, however, because gunpowder rockets had reached their technological limits at a time when the technology of guns and artillery continued to advance.

Even before the nineteenth century began, humans were finally realizing a longstanding dream. Thanks to advances in ballooning, they were taking to the air. On September 19, 1783, Joseph and Etienne Montgolfier launched a large lighter-than-air balloon at Versailles with a sheep, a duck, and a goat as passengers. On October 15, François Pilatre de Rozier, a teacher, became the first man to ascend in a balloon when the Montgolfier brothers let him fly in a series of tests of a tethered balloon. On November 21, de Rozier and the Marquis d'Arlandes made the first free flight in a Montgolfier balloon, flying five miles for twenty-five minutes up to a height of three thousand feet. Jacques Charles, who was experimenting at the same time with hydrogen balloons, made his first ascent with his brother Noel ten days later. The age of ballooning had begun, and over the decades that followed, balloons carried larger loads, traversed longer distances, and reached greater altitudes.

Fakes, Fantasies, and Facts

Early in the twenty-first century, it is a cliché to observe that the twentieth century produced changes that had revolutionized life in ways that people one hundred years before couldn't have visualized. There is no doubt that many people a century ago felt the same about the nineteenth century, thanks to the new modes of transportation, communication, manufacturing, and warfare that hadn't been envisioned in 1800, such as steam turbines, telephones, telegraphs, wireless communications, photography, new steels and textiles, and electrical devices, to name a few.

Technological changes in the nineteenth century took longer to propagate than they do today, however. Many of that century's advances that were known to scientists but not average people were popularized through books, newspapers, and other publications read by growing numbers of people who benefited from the dissemination of education and science. But sometimes, large numbers of people were exposed to fakery and misrepresentation, and the largely unknown realm of space was a popular setting for deception. One of the great hoaxes of all time was played on readers of the *New York Sun* in August 1835, when the newspaper published an account of observations made earlier that year by well-known British astronomer Sir John Herschel, son of Sir William Herschel, the discoverer of the planet Uranus. Herschel at the time was observing in South Africa, out of easy reach from New York, and the newspaper started with this true fact to spin a fantastic tale of Herschel's purported observations of the moon through a giant telescope.

According to the cleverly written *Sun* series, which ballooned the circulation of the paper and was later printed in best-selling pamphlets, Herschel observed a wide variety of plants and animals on the moon, including bison, birds, unicorns, biped beavers, and a race of winged humans who worshiped in temples. All were visible through Herschel's telescope, which, the report claimed, could magnify objects more than six thousand times, ignoring the effects of the earth's atmosphere on earthbound telescopes. The series, which was reported to have been taken from the *Edinburgh Journal of Science*, which in reality didn't exist, was probably written by Richard Adams Locke, a British-educated reporter who worked for the *Sun*. Although many educated people doubted the hoax, large numbers believed it, and it was many months before Herschel himself even found out about the existence of the articles.

In the 1860s, a wave of books of speculative fiction circulated stories that later became known as science fiction. A number of novels about trips to the moon, Mars, and Venus appeared in France in 1865 alone, along with Camille Flammarion's book describing the available literature on space science and speculation, *Mondes imaginaires et mondes réels*. Two books stand out from the crowd in terms of their popularity and influence. Jules Verne, in his books *De la Terre à la Lune* (*From the Earth to the Moon*, 1865) and *Autour de la Lune* (*Trip Around the Moon*, 1867) was able to describe a trip around the moon that barely strained the credulity of even informed readers. His characters were products of their time, and Verne relied on the latest scientific knowledge to inform his novels.

Verne, who became a writer after studying law and working as a playwright and stockbroker, had already made a success of his novels on balloon-borne exploration of Africa, the North Pole, and the center of the earth. He had studied Edgar Allan Poe's 1835 account of a trip to the moon by balloon, *The Unparalleled Adventure of One Hans Pfaal.* In an essay on Poe's work, Verne stated that Poe's work was itself an advance on previous stories of trips to the moon because Poe had applied scientific knowledge to make his account more realistic. Of course, to make the balloon trip to the moon possible, Poe could use scientific reality only in limited quantities.

From the Earth to the Moon is as much a book about Verne's perception of America as it is a speculation on spaceflight. It covers the development of the gigantic cannon *Columbiad* and its projectile, which is aimed at the moon. In Verne's story, the moonshot project is conceived and carried out by the Gun Club of Baltimore, Maryland, a large organization of bored Civil War veterans that is on the verge of dissolution because there's no new war in which to test guns and other artillery. Verne's acerbic description of the Gun Club reflects his horror at the massive amount of blood spilled in the Civil War as well as his admiration for the American spirit of exploration and enterprise.

The club's president, J. P. Barbicane, suggests shooting a projectile to the moon as a project to engage the Gun Club in place of another war, and the effort gains wings, supported by a massive public and international subscription. The project design is settled in four meetings with the assistance of a letter from astronomers explaining the moon's orbit, which allowed a launch date to be fixed. The launch of the projectile takes place less than fourteen months after Barbicane's suggestion. With barely two months to go before launch, a gallant Frenchman named Michel Ardan, modeled on Verne's friend Félix Tournachon, better known as Nadar, volunteers to fly aboard the projectile. The spherical cannonball planned for the launch is replaced by a bullet-shaped aluminum craft that carries three men; two dogs; a number of chickens; air, food and water; and a variety of gear that even today's astronauts can only dream about. On launch day, the 900-foot-long cannon *Columbiad* is ready, built into a hill in central Florida between Tampa and a then obscure place called Cape Canaveral. The drama of the coming launch is not supplied by the technical and organizational problems that are part and parcel of every major technological undertaking today, but rather by an opponent, Captain McNicholl, who proposes a duel with Barbicane before finally being won over and volunteering to join the crew. By the time of launch, millions of people from all over the world are on hand to witness the event, something that also happened a century later when another three-man crew left Florida for the moon, this time for real.

The only means of tracking the projectile is a specially constructed reflecting telescope with a mirror sixteen feet across, roughly the same size as the telescope on Mount Palomar that went into operation in 1948. The launch is followed by several days of cloudy weather, part of a climactic disturbance caused by the gigantic explosion that was needed to launch the projectile. The book ends in cliffhanger fashion with a sighting of the projectile, apparently in orbit around the moon.

The second book, *Trip Around the Moon*, follows the three voyagers on their trip around the moon. Having survived the launch thanks to a system of water cushions, the three travelers pass within twenty-four miles of the moon, but their trip takes place during the full moon, so they see almost nothing of the moon's far side. Without a thought about how they might get back, the three space travelers hope to land on the lunar surface and are prepared to meet Selenites. A near collision with a meteoroid on the way out from earth puts them off course for the moon, however, and on a course that takes them back to a splashdown in the Pacific Ocean, just off the coast of Baja California. The three survive their plunge back to earth and are picked up days later by the U.S. Navy.

The biggest problem with Verne's book is the idea that the projectile or its occupants could ever have been successfully launched. Verne glossed over the problems of building a cannon nine hundred feet long and the impossibility of a cannon of that size successfully sending a projectile to the moon because of air resistance, among other problems. Then there's the problem of air resistance inside the cannon and in the lower parts of the atmosphere that would have held back the projectile. And, of course, no way had been devised to allow passengers to survive a cannon launch. Unlike real space travel, Verne's space travelers experienced weightlessness only briefly, at the points in the journey where earth's gravity balanced out the moon's.

Nevertheless, these two books are often based on fact, and they remain famous for their prescience, such as the Florida launch site and Pacific splashdown. A chapter is taken up with the dispute between Florida and Texas for the honor of hosting the *Columbiad*, a dispute that would be replicated nearly a century later when the Apollo control center and astronaut training facility were set up in Houston, Texas, instead of near the launch facilities in Florida. Prior to the launch, Verne's account also has a squirrel and a cat flown in a test projectile before humans step aboard the moonbound projectile. Ninety years later, the Russians tested their first spacecraft with dogs and the Americans with chimpanzees. Verne's books contain the basics of the trajectory work needed to bring the projectile and the moon together, although Verne's account barely scratches the surface of the computing work that would be required to bring a spacecraft near the moon. Verne's launch speed is correct, and the time his projectile needed to get to the vicinity of the moon, about four days, compares well with the real figure. Finally, the projectile is equipped with rockets to change its course, something that also anticipated reality.

During their voyage, Verne's three travelers speculate about the nature of the moon. Though there is plenty of discussion of life on the moon, most of the facts advanced suggest otherwise, and, indeed, the projectile's crew return to earth without seeing any suggestion of life. Much of Verne's scientific discussion is based on the best knowledge of the time, including the belief that most lunar craters were of volcanic origin, something then widely believed but disproved in the twentieth century. While writing these books, Verne consulted more than five hundred reference books, and his two novels refer readers to actual contemporary scientific works on the moon.

Verne moved on to different adventures in his subsequent novels, but his two books about a trip to the moon are widely credited with helping make possible the real lunar trips a century later by inspiring many of the people who would make those trips possible. Verne's ability as a storyteller and his research work, which enabled his speculations to be raised on a foundation of fact, help *From the Earth to the Moon* and *A Trip to the Moon* stand out from the many other books of his time that speculated on space travel.

A contemporary of Verne's who also advanced space travel was New England writer and social critic Edward Everett Hale, whose story "The Brick Moon" appeared in *The Atlantic Monthly* in 1869. Hale's tale marked the first time that the concepts of the artificial earth satellite and the space station appeared in print. Hale's Brick Moon is a navigational device that helps sailors and others fix their longitude in the same way in which the North Star can be used to fix latitude. Designed to carry humans, the Brick Moon is launched by a set of gigantic flywheels that gain energy over years of turning. In Hale's story, the satellite is launched prematurely and unexpectedly while thirty-seven people are on board. Observers on earth later find the vanished craft in orbit, and its occupants send messages to earth by Morse code, which they communicate by jumping up and down on the brick sphere (which carries an atmosphere on its exterior). The satellite receives messages from earth through the medium of giant black letters laid out on snow and occasional shipments sent up by the flywheels. But because the Brick Moon contains provisions, animals, and plants, it is essentially self-sufficient.

Hale's Brick Moon also anticipated a recurring theme in speculation about space stations and space colonies: the idea that these new settlements would be free of the social problems that beset humans on earth. The inhabitants of the Brick Moon refuse a shipment of books on law and government, for example, because they don't need them. Hale ends his story this way: "Can it be possible that all human sympathies can thrive, and all human powers be exercised, and all human joys increase, if we live with all our might with the thirty or forty people next to us, telegraphing kindly to all other people, to be sure? Can it be possible that our passion for large cities, and large parties, and large theatres, and large churches, develops no faith nor hope nor love which would not find aliment and exercise in a little 'world of our own'?"

An interesting postscript to Hale's story is the fact that some of the first actual earth satellites ever launched were, like the Brick Moon, navigational beacons for military ships. Similarly, in the 1990s, the establishment of the Global Positioning System of navigation satellites became an almost indispensable part of the latest technology. Building space stations, however, introduced problems well beyond the imagination of anyone in the nineteenth century.

While speculations on space travel proliferated, astronomers made observations that fueled new fantasies about the planet Mars. In 1877, when Mars made a close passage to earth, astronomers were looking through telescopes of growing size and quality to try and tease out new details from the ambiguous but changing face it presents to astronomers at the bottom of earth's atmosphere. That year the American Asaph Hall discovered Mars's two tiny moons, Phobos and Deimos, and the Italian Giovanni Schiaparelli sketched linear features on Mars that he called *canali*, the Italian word for channels. Schiaparelli's term was mistranslated into the word *canals*, which suggested that they were built by intelligent beings. In the 1890s, a wealthy Boston astronomy enthusiast named Percival Lowell began observing the Red Planet and writing about his findings. Lowell drew canals on Mars and publicized his belief that these structures were being built and maintained by a civilization battling water shortages. It didn't take long for Lowell's suppositions to fire the imaginations of writers.

In Germany, Kurd Lasswitz wrote an influential novel, *Auf zwei Planeten* (On Two Planets), published in 1897. It followed the then conventional wisdom that Mars was older than earth, thus making it possible for a more advanced civilization to have developed. Lasswitz, a mathematics professor in Gotha in central Germany, wrote that the Martians had had time to solve the problem of space travel, and hence they would visit earth before humans could go to Mars. He suggested that the canals were actually corridors of vegetation through the deserts of Mars. The Martians set up their first station on earth at the North Pole, Lasswitz maintained, and they used a gravity-resistant material to move about space. Using this mysterious substance, the friendly Martians brought humans to their spacecraft hovering above the earth. The German spaceflight pioneer Willy Ley, who was among those who were influenced by Lasswitz's novel, wrote: "If we disregard the marvelous but impossible . . . substance which makes things inaccessible to gravity, Lasswitz presented a solution of the problem of space travel as mathematicians see it."

Another writer of speculative fiction who gained fame at the end of the nineteenth century, the Englishman Herbert George Wells, used a similar gravity-defying material to get humans to the moon in his 1901 novel, *The First Men in the Moon*. Wells, who also wrote *The Time Machine* (1895) and *The War of the Worlds* (1898), among many other novels and stories, stood out from other writers of the time in that he had a scientific education under the zoologist, ethicist, and Darwinist Thomas Henry Huxley, and his works are among the first that people instantly recognize as science fiction today.

In *The War of the Worlds*, Wells took Lowell's speculation about life on Mars a step further with a novel about an invasion of the earth by hostile Martians. They first land near Woking, attack London, and then lay waste to southern England with large machines shooting heat rays. The story was seen by many as an allegorical attack on imperialism, then a popular idea in late Victorian Britain. The Martians were not defeated by terrestrial armies but by the tiny earthly microbes that they were not immune to. Wells's story gained new life on October 30, 1938, when the talented actor and director Orson Welles led his Mercury Theater on the Air in a radio dramatization of *The War of the Worlds*. The broadcast, in the form of radio news bulletins, terrified thousands of Americans who believed that Martians had landed at Grover's Mill, New Jersey, and were attacking major American cities. The story has since been resurrected for several movies.

The First Men in the Moon was less an anticipation of the first trip to the moon, as in Verne's story, than a fantasy piece about space travel and about life on the moon. In this novel, Wells's protagonists, Bedford and Cavor, travel to the moon in a glass-and-steel sphere coated with an substance named Cavorite that can defeat gravity. The moon is populated by intelligent Selenites who are specialized according to their station in life, and their major source of food besides plant life is mooncalves. Wells's Selenites were reminiscent of the devolved humans he meets in the far-off future in *The Time Machine*. *The First Men in the Moon* owed its inspiration to many earlier works, notably Lasswitz's novel and Kepler's *Somnium*.

Wells continued to write, but his later works, influenced by his political, technological, and utopian enthusiasms, have not endured like his early work. Wells died shortly after the end of World War II, when he could see how technology in the form of the atom bomb, the rocket, the bomber, and the concentration camp, had clouded his hopes for the future of humanity.

Another writer who drew inspiration from Percival Lowell and charged imaginations with his visions of other planets was Edgar Rice Burroughs, who is best remembered today as the creator of Tarzan. His prodigious output of books and stories in the first half of the twentieth century also inspired many people to think about travel to Mars and Venus. After a stint in the military and a failed career in business, Burroughs got his start as a writer in 1912 when he sold a story about Mars to a pulp fiction magazine. That same year he also published *Tarzan of the Apes*, the first of many books about his most enduring character. Among his other famous characters was John Carter, who went to Mars by astral projection, and living among the civilized but warlike Martians became the star of many stories and eleven books. Another Burroughs hero, Carson Napier, took up several books with his adventures on Venus. Burroughs paid little attention to how his heroes got to Mars or Venus, but stressed their planetary adventures, which were very popular. Like Jules Verne, Burroughs set other adventures deep inside earth, writing the well-known *The Land That Time Forgot* trilogy. Burroughs's work appeared regularly in the early science fiction pulp magazines along with that of other masters like H. G. Wells, inspiring new generations of science fiction writers and space enthusiasts, among them Ray Bradbury and Arthur C. Clarke.

Even as Wells, Burroughs, and other writers created their speculations on inventions and travel to other planets, new technologies and ideologies also developed in the nineteenth century were being seen as solutions to the problems of humanity. The twentieth century would see these nineteenth-century products used, along with brand-new inventions, in ways that no one could have predicted. Events in Russia, in 1900 a primitive and troubled empire on the margins of Europe, would within a generation propel obscure revolutionaries to power and set off a chain of events that would send humans into space. One of the early revolutionaries whose death inspired the efforts of other revolutionaries also dreamed of flying with rockets. Their political efforts became a big part of the story of humanity's reach into space.

2. Tsiolkovsky and the Birth of Soviet Astronautics

All my works on aviation, rocketry, and interplanetary travel
I bequeath to the Bolshevik Party and the Soviet government,
the true leaders of progress in human culture. I am sure they
will be able to consummate these works successfully.

Konstantin Tsiolkovsky's last will and testament, 1935.

The technological changes that emerged in the nineteenth century brought
with them far-reaching social, economic, and political changes. Growing
populations and industrialization forced millions of people in Europe to shift
from agrarian to urban settings, a transition that was difficult for most. The
American Revolution of 1776 and the French Revolution of 1789 heralded
a period in world history in which political ideology and nation-states replaced religion and royalty as the driving forces behind politics and war.

As the nineteenth century drew to a close, few places were experiencing
change as jarring as Russia and its empire. In 1861, Tsar Alexander II had
ended serfdom, and although the poor peasants of the Russian countryside
suffered under new forms of subjugation, Russia moved from a feudal system to industrialization in little more than a generation. The wrenching economic changes that resulted encouraged revolutionary violence, much as had
the similar agitations that racked Europe during the nineteenth century.

The American Revolution introduced representative democracy, and the
French Revolution took place under the banner of liberty, equality, and fra-
ternity before it collapsed into a bloody morass that ended in dictatorship.
Industrialization and the creation of a large class of urban working people
across Europe led to the ideas of Karl Marx and to parties who called for
workers' control of the economy through socialism and communism. Many

other political currents moved in the century, including conservative movements reacting to socialism, more liberal philosophies representing the values of the emerging middle class, and some extreme utopian philosophies such as anarchism. Nationalism and the idea of the nation-state also moved to the fore during this period.

Although Europe was relatively free of war between the fall of Napoleon in 1815 and the start of World War I in 1914, these new political movements challenged those in power, notably during a series of foiled revolutions in 1848. In the late nineteenth century many revolutionary currents had responded to the changes taking place in Russia, and on March 1, 1881, members of a group called Narodnaya Volya (The People's Will) assassinated Tsar Alexander II by throwing a bomb into his carriage in St. Petersburg. The bombers hoped to end autocracy in Russia and replace it with democracy and socialism, but in the short term they succeeded only in replacing Alexander with a more reactionary and repressive czar, Alexander III.

The maker of the bomb, a twenty-seven-year-old scientist named Nikolai Kibalchich, was arrested and thrown into a tiny cell in the Fortress of Peter and Paul. Knowing that he would be put to death, he set to work on designing an aircraft using reactive devices to move it. Before he was hanged on April 3, Kibalchich wrote a memorandum on his craft, which used a solid rocket to propel a platform through the air. "I believe in the feasibility of my idea, and this faith supports me in my terrible situation," he wrote in his cell. "If learned specialists find my idea realistic I shall be happy to be able to render service to my country and mankind. I shall meet death calmly then, knowing that my idea will not perish with me." He begged authorities to share his memorandum with other scientists, but after his execution the memorandum was hidden in the government archives for decades until the revolution he had hoped for enabled his work to be published. Although his device would not have worked as hoped, it did point the way to the use of rockets as a means of moving humans through the atmosphere and beyond. And Kibalchich became the first of many rocket enthusiasts to be entangled in Russia's knotty revolutionary politics. Few would be as willing.

During this same period, a twenty-four-year-old schoolteacher in the small and obscure Russian farming town of Borovsk, about seventy miles south of Moscow, was settling in to a new marriage and working hard on chemi-

1. Soviet postage stamp honoring Konstantin Tsiolkovsky.

cal theories he thought were new. Konstantin Eduardovich Tsiolkovsky sent his paper, "The Theory of Gases," to the Society for Physics and Chemistry in St. Petersburg that same year and was told that most of his "discoveries" about the kinetic nature of gases had been made by others more than twenty years earlier. The news must have been discouraging, but the arithmetic teacher was already used to dealing with the setbacks of life.

Tsiolkovsky was born in the village of Izevskoye in Ryazanskaya province south of Moscow on September 17, 1857. He was the fifth of eighteen children of a forester, Eduard Tsiolkovsky, who had emigrated from Poland, and of Maria Yumasheva, who came from a family of artisans. The family soon moved to the city of Ryazan, and the young Tsiolkovsky enjoyed a happy and carefree childhood until he was nine, when a case of scarlet fever he contracted while tobogganing led to the loss of most of his hearing. "I felt I was isolated, humiliated—an outcast," he wrote of his deafness. "This caused me to withdraw deep within myself, to pursue great goals so as to deserve the approval of others and not be despised."

Soon after his illness, the family moved north to Vyatka. Tsiolkovsky learned to read and write with the help of his mother and enrolled in the local boys' school. But his mother died when he was thirteen, and soon after the boy dropped out of school. Tsiolkovsky began to read his brothers' textbooks and every book he could get his hands on. Becoming interested in physics and flying, he began designing aircraft.

With the support of his father, Tsiolkovsky set off at age sixteen for Moscow, where he pursued his self-education efforts. He rented part of a room and spent as much as he could of his meager allowance on books and supplies for his experiments. "I remember very well that I had nothing to eat but brown bread and water. For all that, I was happy with my ideas, and my diet of brown bread did not damp my spirits." He performed experiments to confirm his observations and readings and also attended public lectures, listening with the aid of an ear horn he built. But most importantly, he spent as much time as he could in libraries studying mathematics, astronomy, physics, and chemistry.

One library he frequented was operated by Nikolai Fedorov, a philosopher who expounded his own theory of cosmism, which holds that the universe is full of life and that humans need to move into space to advance their material and spiritual development. Indeed, Fedorov believed that humanity would have to expand beyond the Earth because humans would eventually gain the ability to bring the dead back to life. Tsiolkovsky absorbed these beliefs and wrote about them throughout his life. Fedorov also tutored Tsiolkovsky and provided him with books, clothing, and food to make his difficult life more bearable. Tsiolkovsky's imagination was also fired by Jules Verne's novels, which he said "startled my brain." The young student was convinced that humans must take to the air and fly into space as soon as possible, and so he set his mind to reaching these goals.

Tsiolkovsky's spartan lifestyle left him looking sickly, however, and his clothing was worn and marked with acid burns from his chemistry experiments. After three years in Moscow, Tsiolkovsky was called back home by his father. After further studies, Tsiolkovsky passed his teaching examination and took a job in Borovsk. In 1880, he married Varvara Sokolova, the daughter of a local preacher, who became his devoted companion. Although his 1881 chemistry paper had replicated previous results, it was clear to some leading members of the Society for Physics and Chemistry, such as Dimitri Mendeleyev, inventor of the periodic table of elements, that Tsiolkovsky had reached his conclusions independently and had scientific talent. Tsiolkovsky was later admitted to the society when he submitted a second paper on the effects on humans of natural forces such as gravity. In 1883, he wrote a series of diary notes that were published after his death as *Free Space*. These notes contained his thoughts about motion in space free of gravity, including discussions of weightlessness and rocket propulsion, although in these notes he erroneously viewed propulsion as the result of the release of gas from a vessel, which thus caused the vessel to move.

During his years as a science teacher in Borovsk, Tsiolkovsky also devoted a great deal of spare time to pondering questions related to aircraft. He did substantial research on metal dirigibles and constructed many models of these craft, years before the dirigibles and zeppelins of the twentieth century. His first published paper, in 1891, was entitled "The Problem of Flying by Means of Wings," and dealt with what later became known as airplanes. As part of his research into dirigibles, he constructed Russia's first wind tunnel in 1897 and conducted research on metal monoplanes. In 1899 Tsiolkovsky applied for a grant from the Russian Academy of Sciences to support his wind tunnel research; the grant of 470 rubles was the only government support Tsiolkovsky received for his research during the tsarist years.

Tsiolkovsky and his family had moved to the provincial capital town of Kaluga when he won promotion to a teaching job there in 1893. Kaluga remained his home for the rest of his life, and the town's name has become almost synonymous with Tsiolkovsky. It was in Kaluga that Tsiolkovsky completed his most important work, "Investigating Space with Reaction Devices," which was completed in 1898 and published five years later in the journal *Scientific Review*. The year of publication, 1903, is remembered for the Wright brothers' first successful flight of an airplane. Tsiolkovsky's great contemporary achievement was little known because government authorities confiscated most of the copies of that issue, apparently because of some offensive content elsewhere in the journal.

This work, one of the great foundations of astronautics, contained the first proposal for the use of liquid-fueled rockets as devices for venturing into space, including rockets that use the combinations of liquid hydrogen and liquid oxygen, kerosene and liquid oxygen, alcohol and liquid oxygen, and methane and liquid oxygen. Tsiolkovsky proposed that the fuel and oxidizer be mixed together in a combustion chamber, and also suggested that rudders in the rocket's exhaust be used to steer the vehicle. This 1903 article also contained equations that proved that such rockets could carry people into space because liquid fuels, among their other qualities, were potentially far more powerful than the gunpowder-based solid rockets of the time. By computing the power of fuels and the speeds necessary to escape

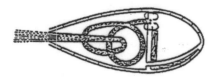

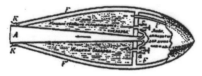

2. By the end of the nineteenth century, Konstantin Tsiolkovsky was examining the fundamental scientific theories behind rocketry. In the 1920s, Tsiolkovsky analyzed and mathematically formulated the technique for staged vehicles to reach escape velocities from earth. (NASA photo)

the Earth, Tsiolkovsky showed that while a single rocket was not up to the job, "step rockets" could succeed. One type of step rocket was a "rocket train," today known as staged rockets, where the bottom rocket stage burns first and then drops off, leaving the next stage to burn. Each stage burns and falls away until the rocket has succeeded in reaching the desired velocity. Although he was not the first to think of a staged rocket, Tsiolkovsky was the first to see its important role in reaching space. (Efforts late in the twentieth century to build single-stage rockets capable of going into orbit fell short because the more powerful fuels and lighter-weight structural materials such vehicles would need to reach orbital or interplanetary velocities were still unavailable.)

The small solid-fuel rockets of the nineteenth century were far too limited to move humans or payloads into space. With his proposal to use liquid-fueled, staged rockets, Tsiolkovsky was the first to show a feasible way to go into Earth orbit and beyond. Tsiolkovsky elaborated on these groundbreaking theories in expanded and updated versions of his 1903 article, and he also looked at some of the human factors of space travel. For example, his notebooks included drawings of spacesuited humans exiting their spacecraft through air locks, and he wrote of the effects of weightlessness in space. He also continued to think about advances in aviation, writing theoretical papers about using jet engines in future aircraft. Tsiolkovsky also wrote about his ideas in novels such as *Dreams of Earth and Sky* and *Outside the Earth.*

Tsiolkovsky continued to teach and, in his spare time, experiment and write. But the shattering of Europe's peace with the beginning of World War I in 1914 permanently altered his life. Russia had been shaken by political violence in 1905, which had been followed by a period of some reform. But in the war the failings of the tsarist system led to Russian military setbacks against its German and Austrian-Hungarian opponents. Early in 1917, a revolution deposed Tsar Nicholas II, and a series of moderate governments attempted to govern the country until another revolution that fall put the Bolshevik communists, under Vladimir Ilyich Lenin, in power. Lenin withdrew Russia from the war, but the country became embroiled in a civil war that went on until 1922, when the Communist Party established the Union of Soviet Socialist Republics. By then, Lenin was in poor health, and the Communist Party was wracked by internal turmoil, which culminated in the rise to power of Joseph Stalin. Stalin brutally ruled the Soviet Union until 1953 through repressive political police, a vast network of prison camps, executions of political prisoners, and famines that killed millions.

Kibalchich was hailed as a father of both the Soviet revolution and of spaceflight, with the implication that both went naturally together. Tsiolkovsky's talks near the end of his life praising the Soviet revolution were adduced as evidence that communism was linked to the progress implied by the exploration of space. But as with the later victories of Soviet science, the fall of the Soviet Union in 1991 has given historians opportunities to look at these Soviet claims more critically in the light of newly public evidence. They show that the Soviet government did not always treat Tsiolkovsky as a hero.

The second Russian Revolution of 1917 took place a few weeks after Tsiolkovsky's sixtieth birthday. In all likelihood, he was still wrapped up in his own work and cosmism beliefs, which didn't accord completely with communist ideology. Recently revealed documents have shown that the communist secret police arrested Tsiolkovsky shortly after the revolution and held him in jail for several weeks. The elderly teacher was fortunate to have a high official intervene on his behalf and arrange his release. In 1919 Tsiolkovsky was then admitted to the Socialist Academy of Science, which later became the Soviet Academy of Science, and granted a pension, which allowed him to concentrate on his research. Beginning in the mid-1920s, after the Russian civil war and international recognition of the work of Robert Goddard in America and Hermann Oberth in Germany, the Soviet government published Tsiolkovsky's scientific writings, and he became an honored and prominent citizen. Some of Tsiolkovsky's science fiction writings were also published, but very little mention was made during his life or after of his belief in cosmism. Although Tsiolkovsky's convictions included human mastery of nature, something that harmonized with communist ideology, Soviet-era biographies of Tsiolkovsky omitted all mention of Fedorov and cosmism. This aspect of his life, along with his incarceration, remained unknown until after the Soviet Union fell in 1991.

In the early 1930s, Tsiolkovsky wrote a number of unpublished essays about his beliefs in the existence of life throughout the universe. These essays show that he gave a great deal of original thought to the possibility of extraterrestrial life. He believed that life was common in the universe and that the same laws applied in all parts of the universe. "There is no substance that cannot take the form of a living being," he wrote. "The simplest being is the atom. Therefore the whole universe is alive and there is nothing in it but life. But the level of sensitivity is endlessly various, and depends upon the combinations of which the atom is a part." In these essays Tsiolkovsky posed the question later famously raised in 1950—years after his death—by physicist Enrico Fermi: If there is intelligent life elsewhere in the universe, why haven't we seen any evidence of it? Tsiolkovsky answered what is today known as the Fermi Paradox by pointing out that more advanced beings have no reason to visit us because we have nothing to give them. "Apparently, there is hope that something worthwhile will develop from us," he wrote, anticipating more optimistic arguments that advanced civilizations have set Earth aside as a sanctuary.

While he was setting down these ideas, Tsiolkovsky was an honored member of the Soviet state. His more practical beliefs about space travel were broadcast from his home in Kaluga to the 1933 May Day celebration in Red Square in Moscow: "For 40 years, I have been working on the rocket motor, but I thought that a journey to Mars could take place hundreds of years later. Time, however, moves quicker, and now I am sure many of you will be witnesses of the first transatmospheric flight."

Indeed, rocket clubs were already being set up in the Soviet Union, and some of the people who would make his dreams come true, such as the young Valentin Glushko, came to Kaluga to visit the old teacher, who died on September 19, 1935. In gratitude for the help he had received from the Soviet state, he bequeathed his papers and other possessions to the state, which responded by building a museum dedicated to his memory in Kaluga.

Soviet-era biographies of Tsiolkovsky made much of the idea that his ideas would have been lost without the support he received from the Soviet government. Tsiolkovsky was isolated by his disability and his residence in small towns, and he was forced to publish many of his five hundred papers on his own, which limited their circulation. After the communist revolution, Russia too was isolated from the rest of the world, and hence much of Tsiolkovsky's work was not widely known internationally until after his death, when others independently arrived at Tsiolkovsky's conclusions about how to go into space. The first copy of a work by Tsiolkovsky reached America before the Russian Revolution, in 1912, but it was not widely available. As historian Frank Winter concluded after investigating the question, "it appears Tsiolkovsky's contributions to the astronautics literature were virtually obscure and inaccessible in the United States during that early period." His name did not become known in the West until the mid-1920s.

Though Tsiolkovsky spent a few weeks in a Soviet jail and had some of his work suppressed, other Russian scientists experienced far greater challenges negotiating the shoals of life in revolutionary Russia. What the man known as Yuri Vasilyevich Kondratyuk could have contributed to space exploration, for example, had he lived under different circumstances, can only be conjectured. As it was, Kondratyuk made huge contributions to astronautics with his works *To Those Who Will Read in Order to Build* in 1919 and *The Conquest of Interplanetary Space*, which was published in 1929. His best-known idea was lunar orbit rendezvous, which was adopted by the United States and the Soviet Union for their manned lunar landing programs in the 1960s. Instead of sending a craft straight to the moon and returning directly, a crew of astronauts in a mother ship would be sent into lunar orbit, drop a landing crew to the lunar surface in a smaller specialized landing craft, and then return to the mother ship in the landing craft. Finally, the mother ship would leave lunar orbit and return to Earth. Though this method involved complicated rendezvous and docking maneuvers in lunar orbit, it vastly reduced the total weight of the spacecraft being sent to the moon's surface and reduced the complexity of spacecraft design.

Kondratyuk also proposed slingshot maneuvers to speed a spacecraft from one planet to another, an idea that today is used regularly in planetary exploration. He also conceived "aerobraking" as a way to lower a craft's orbit above a planet's surface in a planetary atmosphere, as well as atmospheric braking for spacecraft descending to a planet's surface. Orbital refueling for spacecraft, space mirrors for lighting up the ground, solar panels, fuel cells, electric propulsion, and nuclear-powered generators for spacecraft, among other ideas, were all first proposed by Kondratyuk. Independently of Tsiolkovsky, Kondratyuk also wrote about staged rockets and liquid hydrogen as a rocket fuel.

Clearly, these ideas, which stand only behind Tsiolkovsky's in their importance for space exploration, should put Kondratyuk in the first rank of spaceflight theorists, yet his contributions are not widely known. A major reason for this is his strange and difficult life story, which was intertwined with the political storms that swept Russia in his time. To start off with, Kondratyuk was not his real name. In fact, he was born Alexander Ignatyevich Shargei on June 21, 1897, in Poltava, Ukraine. Despite losing both his parents before his thirteenth birthday, he excelled in school and earned an opportunity to enroll in the St. Petersburg Technical School. Given that his first major work was published when he was only twenty-one, Shargei must have gained his interest in space travel early on. His biographers suggest that he may have read Yakov Perelman's 1915 compendium of spaceflight theory, *Interplanetary Travel*.

Shargei's studies at the technical school were cut short late in 1916 when he was inducted into the tsarist army, trained as a junior officer, and sent to duty on the Turkish front. After Lenin's communists took power and ended Russia's participation in the war in 1917, Shargei was conscripted into the White Army, the forces opposed to the communists during the Russian civil war. On two occasions he deserted the White Army, and after the second attempt he went into hiding, fearing the possibly fatal consequences of being caught by the White Army, from which he had deserted, or the communist Red Army, which would suspect him because of his White Army service. During this period Shargei was able to write his first major work on spaceflight theory, which contained his original ideas on interplanetary travel.

Kondratyuk and Tsander

Shargei's life was still in danger even as the civil war wound down, but when his stepmother learned that a young man named Yuri Kondratyuk had died of tuberculosis in March 1921, Shargei, then back home in Ukraine, assumed Kondratyuk's identity that August. He later moved to Novosibirsk in Siberia. In 1929, he published his masterwork, *The Conquest of Interplanetary Space*, which added to his earlier work and benefited from his further reading and correspondence with Tsiolkovsky and others. Kondratyuk was arrested in 1930 for anti-Soviet activities, a charge that was becoming increasingly common in Stalin's paranoid communist state. Although Kondratyuk was found guilty of economic sabotage, he avoided banishment to a work camp in the gulag and was allowed to work under house arrest in Novosibirsk. He was released in 1932 and got a job building energy stations in the Crimea. His work often took him to Moscow, and rocket enthusiasts there attempted to recruit him, but he declined. In June 1941, Adolf Hitler's German army invaded the Soviet Union, and Kondratyuk was soon back in uniform. He died in fighting near Kaluga in the defense of Moscow between February 22 and 25, 1942.

Kondratyuk's life contains several mysteries. One is his use of the name Shargei when he first published his theories in 1919 and the name Kondratyuk when he again published ten years later. Kondratyuk must have declined the offers to join Moscow's rocketeers in the 1930s for fear that by exposing himself to the identity checks that would accompany work on rocketry, which was of military interest, he would expose his real identity. But this begs the question of how much the Soviet secret police knew about him when he was in their hands in the early 1930s. The Stalinist repression that led to his arrest in 1930 grew far worse in the late 1930s, yet Kondratyuk avoided being swept up in what is now known as the Great Terror. And though the Kondratyuk who died in 1921 was born in 1900, obliging the second Kondratyuk to give that as his birth date, Soviet publications as early as the 1960s gave the space pioneer's birth year as his actual one: 1897. Though Kondratyuk began to get some of the credit he was due in the 1960s, it was only after the fall of the Soviet Union that his life story became known.

Another great Russian space pioneer also met a premature death. Were it not for this fact, Friedrich Arturovich Tsander might be known today as one of the greatest rocket pioneers. Tsander was born on August 23, 1887, in Riga, Latvia, when that Baltic nation was part of the Russian empire. His father, a merchant, raised him after his mother died when he was only two. Even before he left high school, Tsander had been exposed to Tsiolkovsky's writings on space exploration, and the young man dedicated his life to furthering the cause of space travel. After graduating from the Riga Polytechnical Institute, he moved to Moscow and worked in rubber plants during World War I. In 1919, he got work at the Motor Aircraft Factory in Moscow, and two years later, he published his article "Flights to Other Planets," in which he proposed a combination rocket-aircraft for the launch phase of the spacecraft he envisioned. Tsander also proposed that the craft's metallic parts be burned in the rocket's combustion chamber once they were no longer needed, thus creating a means of converting dead weight into fuel. In 1930, he began work at the Moscow Institute of Aviation Machine Building, where he built and tested a jet engine. About this time, he also became involved in a new organization dedicated to building rockets.

By then, a number of Russians were working on rockets. Even before the revolution, a chemical engineer named Nikolai Tikhomirov worked on a solid-fueled projectile, but his ideas didn't win government approval until after the revolution. In 1921, the military set up a special laboratory for Tikhomirov and his assistant Vladimir Artemev. The laboratory began in Moscow but soon moved to Leningrad, as St. Petersburg was called during Soviet times, and grew to become the Gas Dynamics Laboratory (GDL). This group concentrated on building solid-fuel rockets for military use.

While Tsander worked hard designing rockets, perhaps his biggest contribution came as a propagandist for space travel. He spoke to anyone who could help, including Lenin, the founder of communist Russia, who is said to have listened sympathetically. After Tsander lectured on flights to other planets on January 20, 1924, before an audience of Moscow amateur astronomers, the Society for the Study of Interplanetary Communications was formed, with Tsander as an active officer and Tsiolkovsky as an honorary member. The group held several meetings and gained nearly two hundred members. It disbanded late that year, however, probably due to poor finances, after a large meeting was held to discuss what turned out to be a false rumor about American rocket pioneer Robert Goddard sending a rocket to the moon. Other groups in the Soviet Union shared the interest,

3. The original members of the Group for the Study of Reactive Propulsion (GIRD, the Russian acronym) are shown here in 1932. Top center is Yuri Pobedonostsev; bottom center is Sergei Korolev. On the extreme right is Friedrich Tsander, the space visionary who founded the group. (NASA photo)

however. From April to June 1927, the Interplanetary Section of the Association of Inventors held the world's first exhibition of "interplanetary vehicles" in Moscow, which included contributions from space enthusiasts in the Soviet Union, Germany, France, Austria, America, and elsewhere.

While the GDL continued to grow and conduct work on rockets fueled by gunpowder and other solid fuels, its members also did a small amount of work on liquid-fuel rocket engines using various fuels. They even crafted the world's first electric rocket engine in 1929. GDL employees and other Leningrad rocket enthusiasts, which included a young engineer from Ukraine named Valentin Petrovich Glushko, began testing rocket engines behind the walls of the Peter and Paul Fortress, where Kibalchich had spent his last days writing his memorandum on rocket flight. Glushko was born and raised in Odessa, the son of a Ukrainian father and a Russian mother. He was only fifteen when he first wrote Tsiolkovsky in 1923 to declare his devotion to the idea of space travel. Although trained as a sheet-metal worker, he began experimenting with unexploded shells left around Odessa. When he couldn't find a space-related field of study at Leningrad State University, Glushko left in 1929 and joined GDL, where he began his life's work on rocket engines.

In 1931, rocket enthusiasts in Moscow and Leningrad founded similar research groups. In the Soviet state, these groups were strictly regulated, and both the Leningrad and Moscow groups were called the Group for the Study of Reactive Propulsion, known by its Russian acronym, GIRD. Members of GIRD tended to be heavily focused on building rockets, usually with liquid fuel, for space travel. The next year, the Leningrad GIRD group amalgamated with GDL, which maintained its military and solid rocket orientation. In the Moscow GIRD, Tsander headed one of four experimental and research teams, and another was headed by a young glider pilot from Ukraine named Sergei Korolev. In addition, rocket enthusiasts elsewhere in the Soviet Union formed GIRD groups, but little is known about them. Some GIRD members conducted research, and others, such as Yakov Perelman and Nikolai Rynin, publicized space exploration and rockets. Perelman wrote a bestseller, *Interplanetary Travel*, along with popular science fiction stories, and he helped found the GIRD group in Leningrad. Rynin, an engineering professor, wrote the nine-volume encyclopedia *Interplanetary Flight*, published between 1928 and 1932. Because Rynin and Perelman were well known through their writings, they also acted as liaisons between the Soviet rocketeers and spaceflight enthusiasts in other countries.

The Moscow GIRD teams worked in a laboratory in a stone cellar of a building in central Moscow near the U.S. embassy. There Tsander exhorted his comrades with the shout "Forward to Mars!" and talked often of his desire to visit the Red Planet because it has an atmosphere and could harbor life. He spoke of the other planets too, and underlined his enthusiasm by naming his son Mercury and his daughter Astra. Tsander was known to neglect his appearance and even to forget the ration card that was a prerequisite to eating. He and his GIRD comrades worked around the clock through 1932 on a new rocket engine that they hoped to attach to a flying wing. Both the aircraft and Tsander's engine were plagued with problems, however. But the GIRD teams kept working on the engine, even as exhaustion forced Tsander to a sanatorium in the Caucasus Mountains. His comrades test-fired his

engine in his absence even as news came back that Tsander had caught ty-
phus and was very ill. He died on March 28, 1933.

Korolev's Early Life

The loss of Tsander put Sergei Pavlovich Korolev into a leadership role in
GIRD and eventually into a position to make many of Tsander's dreams
come true. Born on December 30, 1906, in the city of Zhitomir near Kiev,
Korolev was the son of a Russian father and a mother from an old Cossack
family whose unhappy marriage fell apart when Korolev was three. He
and his mother, Maria Nikolaevna Moskalenko, moved to another nearby
town, Nezhin, to be with her parents, and Korolev grew up believing his
father, Pavel Korolev, a teacher, had died. Although this wasn't true, Korolev never saw his father again. The boy was raised largely by relatives, as
his mother spent a great deal of time in Kiev taking courses. No other children lived nearby. "I did not have any childhood," Korolev later said. At
age six, he fell in love with aviation when he saw the Russian barnstormer
Sergei Utochkin fly his aircraft at Nezhin. In 1916, as World War I brought
troubled times to Russia, Korolev's mother married an engineer, Grigory
Balanin, who moved the family to Odessa the next year when he got a job
running the Black Sea port's power station. In 1918, in the wake of the October Revolution that brought the communists to power and led to civil
war, Odessa became one of the most hotly contested places in the former
Russian empire. After a brief Soviet takeover, the town was occupied by foreign troops, which was followed by war between Soviet and counterrevolu-
tionary troops that brought danger, hunger, and disease to its inhabitants.
The final Soviet victory in Odessa in 1920 did not end the hunger and pri-
vation for some time.

Korolev attended a vocational school where he learned to be a tile layer
and roofer. One of his biographers, James Harford, has described the school
as an unusual one in which some of the nation's best physics, mathematics,
and arts teachers inspired their pupils. The teenager also kept up his interest in aviation by helping out at a local military seaplane base and joining
a local aviation club. He persuaded pilots to take him flying, and he began designing gliders. In 1924, Korolev enrolled in the aviation section of
Kiev Polytechnical Institute, where he learned the basics of aeronautical
engineering and took a special course in gliders. After two years in Kiev,

he enrolled in the Moscow Higher Technical School, and took more advanced courses. After his graduation as an engineer in 1929, Korolev used his spare time to pursue his interest in gliders, and those he helped design did well in national competitions. He built a glider at the Central Aerohydrodynamic Institute under the supervision of Andrei Tupolev, who was on his way to becoming one of the Soviet Union's most famous aircraft designers. Korolev got work in the aviation industry and a pilot's license, but his career was interrupted when he contracted typhoid fever and required months to recover. After recuperating in 1931, he married Xenia Vincentini, whom he had courted since their first meeting at the vocational school in Odessa years before.

Where Korolev first came in contact with the idea of spaceflight remains unknown, although Harford suggests that he would have met spaceflight enthusiasts while attending school in Moscow. He may have even attended the 1927 exhibition on interplanetary vehicles. Though accounts exist of Korolev meeting Tsioklovsky, Harford argues convincingly that they never met. In 1929, while he was working on gliders at the institute, Korolev met Tsander, and soon began helping him with his work on rocket engines. Before too long, Korolev was volunteering at GIRD, which took more and more of his time.

Although a stocky man, Korolev was a skilled gymnast in school, and he was known to be brave and a strong leader. But he was also argumentative and prone to use strong language to make his points. Yaroslav Golovanov, his Russian biographer, wrote: "He was not devious, but he was skilled in reaching his objectives and at making necessary compromises." In the Soviet Union, such skills were vital. Korolev was also known to be driven to reach his goals, and he was not afraid to hurt others in their pursuit.

A few months after Tsander's death, on August 17, 1933, Korolev and his comrades at GIRD tasted success, when a new rocket engine fueled by gasoline and liquid oxygen was used to launch the Soviet Union's first liquid-fuel rocket at a testing ground west of Moscow. The rocket was designed by Mikhail Tikhonravov and built by Korolev's team. Another rocket built by Tsander's former team flew in November.

The Moscow GIRD's work attracted the attention of Gen. (later Marshal) Mikhail Tukhachevsky, the forward-thinking chief of armaments in the Red Army, who was already responsible for the GDL in Leningrad. He provided support for GIRD, and in 1933 he merged the GDL and GIRD into a new organization based in Moscow called the Reactive Scientific Research Institute or RNII, which worked on both liquid and solid rocket engines as well as on solid rocket projectiles. Then, Tukhachevsky placed the organization under Ivan Kleimenov, an experienced military engineer, with Korolev as his deputy.

In a recent paper on the troubled history of the institute, Russian space program historian Asif Siddiqi has written that although both the members of GIRD and GDL supported the merger, the marriage of the two groups was rocky. Tukhachevsky was soon shifted to other work, so he exercised little oversight of the institute. Kleimenov favored the military work with solid rockets at the expense of the liquid rocket work of the former GIRD members, which caused Korolev and others to be demoted or fired from RNII if they didn't resign. The engineers at RNII also clashed over whether to build winged rockets and whether the limited liquid rocket research should focus on using nitric acid or liquid oxygen as the rockets' oxidizer. Siddiqi wrote: "The three technical disagreements within RNII stemmed from the two factions' differing visions of rocketry's future. One sought to satisfy short-term military imperatives; the other aspired to the heavens."

In spite of the RNII's low status after Tukhachevsky's shift and the move of RNII to another ministry, Kleimenov did advance the institute's military rocket work, so much so that it resulted in the Katyusha solid-propellant rockets used as artillery in the World War II. Korolev was replaced as deputy by Georgi Langemak, and the demotion may have saved Korolev's life. Kleimenov's decisions about which projects and individuals to advance led to rivalries that in the Russia of the late 1930s had fatal consequences. Technical differences were turned into political struggles. In 1937, the Central Committee of the Communist Party passed a resolution at Stalin's urging calling for stepped-up action against "enemies of the people," inaugurating a new phase in the Great Terror. In May, Tukhachevsky and other top officers were arrested and executed on accusations that they were spies, the result of rivalries within the higher ranks of the military. "Almost all who worked on a project discussed with or authorized by him, or were in contact with him—as all leading rocket specialists were—had now to face the danger of being proclaimed an 'accomplice of a spy,'" said Grigory Tokayev, a military engineer who later defected to the West.

Two engineers who had lost their jobs denounced Kleimenov, who in turn was forced out as head of RNII. He was later arrested in November 1937 and executed the following January following a twenty-minute trial. Langemak was also executed, as were hundreds of thousands of middle managers in related ministries. In March, Glushko was seized, and Korolev, in his turn, received the dreaded visit from the secret police in June. Both Glushko and Korolev were sentenced to terms in the gulag, as were many other rocket engineers of the time. Though some denunciations were sparked by envy or to gain revenge, others were extracted in the torture chambers of the Soviet secret police. In many cases, a single denunciation led to arrest and incarceration or death. The madness of Stalin's purges deprived Russia of its best political, military, and scientific talent just as war clouds were gathering over Europe. The survivors of the purges at RNII completed work on the Katyusha, which was the only rocket weapon available to the Soviets in the struggle with Germany.

Though it has become conventional wisdom that the Great Terror drastically changed the direction of Russian rocket research on the eve of war, Siddiqi has brought this belief into question. Agreeing that the removal of people such as Korolev and Glushko from rocket research for most of the war was important, he notes, however, that recent evidence from Soviet archives shows that the RNII was concentrating on small solid-fuel military rockets even before the purges of 1937 and 1938. "Debates prior to the arrests had already established technological priorities that the purges did not radically alter," Siddiqi maintains. He argues that the Katyushas, which were inexpensive and effective when launched in large volleys from their portable launch platforms, "proved far more effective and efficient than the German V-2."

Prior to the Great Terror, Korolev and his colleagues at GIRD and RNII were grappling so intensively with the prosaic problems of rocket engines that their hopes of rocket flights and space travel appeared a distant dream. In 1936, Korolev wrote: "the time will come for the first terrestrial ship to leave the Earth. We probably won't live to see it, and are destined to spend out life pottering about here below." Two years later, as Korolev began his odyssey through the jails, interrogation rooms, and work camps of the dreaded secret police empire known as the gulag, even his 1936 assessment of the future must have seemed wildly optimistic.

3. Robert Goddard's Solitary Trail

We learned something today. We won't make this mistake again.
We'll correct it.

Robert Goddard's characteristic reaction
to the failure of a rocket test.

The 1927 exhibition of "interplanetary vehicles" in Moscow gave prominence to the work of an American physics professor named Robert Goddard. The grateful Soviet space enthusiasts sent Goddard a scrapbook of the exhibition, but Goddard and his wife Esther felt that the exhibition gave Goddard insufficient recognition for his accomplishments. The Soviets lavished praise on someone Goddard had never heard of, Konstantin Tsiolkovsky, describing him as the father of their rocket program. Goddard had a similarly reserved attitude to rocketeers in Germany, who he felt were stealing his ideas, and even to other space enthusiasts in the United States, whose experiments in rocketry wouldn't wait for Goddard's blessing.

By 1927, Goddard was famous in the United States and around the world as the scientist who pointed to the rocket as the means for reaching outer space. Goddard first came to wide attention early in 1920 after the Smithsonian Institution published his landmark paper, "A Method of Reaching Extreme Altitudes," which included the suggestion that a rocket could be sent to the moon. Before this, no responsible scientist had ever seriously or so publicly suggested a feasible method for flying to the moon and beyond, nor had such an assertion been tied to a famed institution like the Smithsonian. Though the writings of Jules Verne and others had fired many imaginations, it was well known that their proposed methods of space travel wouldn't work. Newspaper accounts of Goddard's paper probably gave most people their first indication that flying to the moon was practically possible. Al-

4. Dr. Robert H. Goddard, America's great pioneer of rocket development, shown with one of his rockets. (NASA photo)

though Tsiolkovsky was the first to think of liquid rockets and their use in space travel, his work was not well known inside Russia until the mid-1920s and virtually unknown outside the isolated communist state.

During the 1920s and 1930s, Goddard was one of the most famous scientists in the world. Unlike Tsiolkovsky, whose work on space travel was purely theoretical, Goddard was actually building rockets. The American's speculations about flying to the moon sparked periodic rumors that he would soon attempt to put his theories into practice, such as a hoax in 1924 that he had actually launched a moon rocket, a rumor that caused much excitement among space enthusiasts in Russia. Despite the different circumstances of their lives, however, Goddard and Tsiolkovsky shared many similarities. Both their lives, for example, were shrouded by legends designed to make personal, political, and nationalistic points, and for both of them the truth behind these legends has only recently been uncovered.

Robert Hutchings Goddard was born twenty-five years after Tsiolkovsky, on October 5, 1882, in the industrial town of Worcester, Massachusetts. His father was a bookkeeper who enjoyed a moderately successful career, but Goddard's mother was diagnosed with tuberculosis when her son was a teenager. Goddard was held back in school due to his own illness, possibly exaggerated in the minds of his mother and grandmother. Nevertheless, as a child he was curious about the world around him, and quickly developed interests in science and invention. The young Goddard loved to observe and experiment, but he also enjoyed Jules Verne's books about flying to the moon. In 1898, a Boston newspaper carried H. G. Wells's *War of the Worlds*, and Garrett P. Serviss's *Edison's Conquest of Mars*, and the serialized stories made a huge impression on Goddard.

Two weeks after his seventeenth birthday, Goddard climbed a cherry tree behind the barn of his family's home in Worcester. "I imagined how wonderful it would be to make some device which had even the possibility of ascending to Mars," he later wrote, "and how it would look on a small scale, if sent up from the meadow at my feet. . . . It seemed to me then that a weight whirling around a horizontal shaft, moving more rapidly above than below, could furnish lift by virtue of the greater centrifugal force at the top of the path. In any event, I was a different boy when I descended the tree from when I ascended, for existence at last seemed very purposive." For the rest of his life, Goddard celebrated October 19 as "Anniversary Day" and kept souvenirs of the occasion and photographs of the tree.

Goddard's health problems meant that he entered high school late, at age nineteen. During this time, he wrote essays about space exploration and continued to experiment. When he completed high school in 1904, he was chosen to address the graduating classes of Worcester's three high schools. He argued that scientists were on the verge of finding signs of life on the moon and Mars and would soon develop a perpetual motion machine. Although these ideas would prove false, Goddard wrapped up his speech in memorable fashion: "Each must remember that no one can predict to what heights of wealth, fame or usefulness he may rise until he has honestly endeavored, and he should derive courage from the fact that all sciences have been, at some time, in the same condition as he, and that it has often proved true that the dream of yesterday is the hope of today and the reality of tomorrow." The final phrase became the most famous words of Goddard's life.

Goddard studied physics at Worcester Polytechnic Institute, and then pursued graduate studies at Clark University in Worcester, where he obtained his Ph.D. in 1911. His studies included work on the conduction of electricity in powders, the production of gases by electrical discharges in vacuum tubes, aircraft design, and crystal rectifying devices and vacuum tube oscillators for radios. He was also greatly interested in building high-speed trains that moved through tunnels along magnetic rails. In 1909, as he was entering Clark, Goddard decided that the way into space was by rocket. He had considered many other ways of doing the job, including charged particles, electricity guns, and magnetism, but he decided to concentrate on rockets as the means for realizing his dreams of space travel because they seemed the most practical means of reaching space.

Goddard got work as an instructor at Worcester Polytechnic, and in 1912–13 he was employed as a researcher at Princeton University in New Jersey. Goddard came down with tuberculosis, which cut short his stint at Princeton, however, and returned home to Worcester and got work at Clark University. Over the years, he was promoted to professor and became head of Clark's physics department, maintaining his association with the university until almost the end of his life. Still recovering from his illness in 1914, he began work on a solid-fuel rocket in which fuel was loaded into a firing chamber much as bullets are fed into a machine gun. Goddard began applying for patents covering his ideas on rockets as well as his other work, such as in radio technology. He also did a great deal of theoretical work on liquid-fueled rockets, on the speeds necessary to leave the earth, and on the manner in which a staged rocket, then known as a step rocket, could reach these speeds. Much of this work unknowingly duplicated the findings of Tsiolkovsky. Goddard's efforts to get funding for his rocket research paid off in the form of a modest grant from the Smithsonian Institution in Washington.

When the United States entered World War I in 1917, Goddard sought out military support for his rocket research. Shifting his research temporarily to a location near the Mount Wilson observatory in California, he developed small tube-launched solid-fuel rockets for use by the army, but less than a week after he demonstrated his rockets to the military in November 1918 the war ended. Goddard's rockets were the precursors of the bazooka shoulder-launched antitank rocket used in World War II, which was developed by his graduate student Clarence N. Hickman.

When the war ended and with it his military funding, Goddard returned to Clark and began looking for new sources of financial support for his rocket research. By 1919, his frustration with the limitations of multiple-charge solid rockets caused him to start shifting his research to liquid fuels. It is at this point that Goddard's legend begins to seriously depart from the real facts of his life.

Goddard the Moon Man

In 1961, G. Edward Pendray, himself an important American rocket pioneer, set out the Goddard legend this way:

In January 1920 the Smithsonian Institution got around to publishing the paper Goddard had submitted to it some time before, his treatise on "A Method of Reaching Extreme Altitudes." Dated as of the previous year, this rather remarkable essay on rockets and their possibilities is one of the classic documents in astronautics. Goddard recounted in it the results of some of his early experiments, explained rocket principles, explored basic rocket mathematics, examined the theory of the step rocket (which he had previously patented), and presented some computations to indicate what great heights a rocket could reach if it had an efficient engine and suitable propellants. In this last suggestion he showed that even a rather modest rocket, in theory at least, could go as far as the moon. He suggested that a pound or two of photographic flash powder might be carried by such a rocket to signal its arrival on our satellite.

The newspapers of 1920 made little or nothing of Goddard's mathematics, experiments, or technical contributions—but they had a field day with his suggestion about a shot at the moon. In addition to deriding him as the "moon man," some challenged his scientific knowledge; worse, a few even suggested that he had deliberately distorted scientific facts to make his point. A writer in The New York Times *scornfully remarked that Goddard "only seems to lack the knowledge ladled out daily in high schools." The effect of this derisive copy on the still very shy and sensitive New Englander was to make him shun publicity all the rest of his life, to be wary in describing his work, and reluctant to grant interviews. . . . Thereafter for a while public interest in rockets declined. Goddard was free to continue his experiments without embarrassment.*

The truth about the reaction to Goddard's paper, as analyzed in a 2003 biography by David A. Clary, was somewhat different. In March 1919, when Goddard was looking for new funding sources, he first won fame when he gave an interview to a Worcester newspaper about his rocket research. The story was picked up by newspapers around the United States and overseas. Much of his sixty-nine-page paper had been written before 1914 while he was looking for funding, and its focus was on his multiple-charge solid rocket. When it was published under the obscure imprint of the "Smithsonian Miscellaneous Collections," the Smithsonian sent out a press release in January 1920 announcing Goddard's work. The release, which stressed the research possibilities of the rocket in earth's atmosphere but mentioned the moon flight only as an "interesting speculation," was picked up by the Associated Press. The result was worldwide fame for Goddard, which was stoked by occasional stories that his moon rocket would be launched soon. Although he usually tried to keep his comments restrained and factual, in 1921 he told the *New York Times* that he would soon have ready a "workable model" of the rocket that would go to the moon.

In Clary's words: "Goddard freely granted interviews to newspapermen over the years, aiming to replace sensationalism with scientific reason. He was dull copy in his own words, but exciting stuff when a reporter's imagination ventured into space with him. . . . He hit the lecture circuit, playing on his notoriety to drum up cash." Goddard did restrict his contacts with reporters to those he trusted, and he gave interviews only at times that suited him. His measures to control what was said about him fed the legend that he avoided reporters.

As for the famous *New York Times* editorial commentary of January 13, 1920, Clary notes that Goddard "took the statement in his stride" because he had seen scientific errors in the press before. The *Times* commentary questioned Goddard's assertion that rockets would work in the vacuum of space because it assumed rockets need something to push against, in this case, earth's atmosphere. Goddard took to heart Newton's law that every action has an equal and opposite reaction, and a significant part of his paper contained the first proof that rockets work in a vacuum, contrary to the conventional wisdom of the time. This important part of his paper wasn't mentioned in the Smithsonian press release. Goddard was well known around the Clark campus for his demonstrations of how a gun loaded with a blank twirled when fired in a vacuum, proving Newton's law and disproving the assertion made by the *Times*. Later in life, Clary notes, Goddard did complain about the *Times* article on at least two occasions, but his correspondence shows that he had a good relationship with the newspaper. In any case, though the commentary criticized Goddard, it also compared Goddard to Jules Verne and noted that Goddard's rocket "is a practicable and therefore promising device." Still, the legend of the *Times* commentary lives on, fueled in part by the *Times*'s famous "correction" forty-nine years later in July 1969 when the *Apollo 11* astronauts were flying to the moon. After the correction appeared, Goddard's wife Esther wrote the *Times*: "During [Goddard's] lifetime, and steadfastly since, you presented him and his work in an objective and favorable light. Such publicity over a period of many years has carried far more weight than the small error of 1920."

Goddard was a very practical New Englander who was raised in an era when individual inventors such as Thomas Edison, Alexander Graham Bell, and Orville and Wilbur Wright took pains to ensure that their claims to invention were accepted over the assertions of others. Goddard was happy to publicize his work when he felt the time was right, and he enjoyed sharing his findings with scientific colleagues who were not, in his opinion, competing with him. But he was very guarded around those he felt might be working in the same area as he, including members of the rocket clubs that sprang up in the 1920s and foreign scientists such as Robert Esnault-Pelterie in France and Hermann Oberth in Germany. When Oberth claimed to be the first to think of liquid propellants because he had come up with the idea in 1912, Goddard asserted that he had first thought of them three years earlier. In the event, both had been beaten to the punch by Tsiolkovsky. Goddard took out many patents on his work in an effort to assert his control over the findings of his research and to profit from them.

Regardless of what Goddard thought of the uproar surrounding his 1919 paper, the publicity showed the world that the way into space was now open and that the rocket was the means. Thousands of people who had been inspired by the fiction of Verne, Wells, Burroughs, and other writers began to experiment and form rocket societies to realize their dreams. According to historian Frank H. Winter, prior to Goddard, the rocket was a small "pasteboard amusement device" with limited use elsewhere. "Now, astonishingly and suddenly, it was transformed into a revolutionary way to penetrate space."

The credibility of the January 1920 announcement of Goddard's work was enhanced, Winter has noted, because it came from the Smithsonian Institution, a famous center of scholarship, and involved a legitimate scientist. Winter shows how newspapers spread the word about Goddard and rockets around the world in a time before radio had become widespread. The newspaper publicity was picked up by magazines, silent films, and works of science fiction. A character based on Goddard even appeared in the popular comic-strip space adventures of Buck Rogers.

During the 1920s, with financial support from Clark University and the Smithsonian Institution, Goddard turned away from his multiple-charge solid rocket work and began to grapple with the difficult scientific problems related to liquid-fueled rockets. His rockets used gasoline and liquid oxygen, the latter of which was available in only limited quantities and was difficult to handle because it must be kept at low temperatures. The fuel and oxidizer had to be pumped into a combustion chamber where they were mixed and ignited. Pumps and igniters were complicated, and combustion chamber efficiency varied with its shape. As well, Goddard and those who followed him had to deal with the difficult problem of preventing combustion chambers from melting or burning through.

In 1924, Goddard married Esther C. Kisk, who not only gave him strong support during his lifetime (and afterward), but helped directly with his research by taking notes and photographs of his work.

Goddard made his first firing of a liquid-fueled rocket engine in a laboratory test stand on November 1, 1923. By abandoning heavy pumps to drive the fuel into the combustion chamber and replacing them with a lighter inert gas system, he made progress toward the first practical liquid-fueled rocket. In December 1925, a rocket confined to a test stand tried for a few seconds to lift itself against its restraints. Goddard and his assistants built a flight model of the rocket, and on March 16, 1926, they took the rocket and its launching stand out to his Aunt Effie Ward's farm in nearby Auburn, Massachusetts. This ten-pound, ten-foot-long rocket differed from virtually all the rockets that followed it, in that the engine was built in a pipe frame above the fuel tanks in a "tractor" design. The launch took place at 2:30 p.m. under clear skies.

As Esther Goddard filmed the scene, assistant Henry Sachs lit the igniter, and Goddard turned a valve. "Even though the release was pulled, the

5. Dr. Goddard with his complete rocket with double-acting engine in November 1925, following more than two years of development of the idea of separate pumps for each propellant. Dr. Goddard's first successful liquid-propellant rocket shot took place on March 16, 1926. (NASA photo)

rocket did not rise at first, but the flame came out, and there was a steady roar," Goddard wrote the next day. "After a number of seconds it rose, slowly until it cleared the frame, and then at express train speed, curving over to the left, and striking the ice and snow, still going at a rapid rate. It looked almost magical as it rose, without any appreciably greater noise or flame, as if it said, 'I've been here long enough; I'll be going somewhere else, if you don't mind.'" The rocket rose 41 feet and landed 184 feet away from its launch stand after two and a half seconds of flight.

"This flight, to be sure, was small and was not considered by the Smithsonian Institution as a flight that would persuade the public of the possibili ties of the method," Goddard wrote later. "As a first flight it compared with the Wrights' first airplane flight, however, and the event, as demonstrating the first liquid-propelled flight, was just as significant." Goddard waited a few weeks to report his success to the Smithsonian and said nothing to the media. Indeed, he didn't publicly acknowledge this launch for another ten years, unlike the Wright Brothers, who informed the press shortly after their historic flight. He also kept his findings under wraps because of his concerns about reports of rocket building in Germany.

Goddard worked on improvements to his design, including moving the engine to the bottom of the rocket, but his later flight attempts failed for a variety of reasons. He pronounced each a learning experience. The following year, reacting to the pressure from the Smithsonian to fly longer and higher, Goddard built a much bigger rocket that weighed 150 pounds, but it exploded during a test. He then built a scaled-down rocket, though it was still larger than the first 1926 rocket, and this one flew in December 1928.

On July 17, 1929, Goddard was back at Aunt Effie's farm with a new rocket equipped with a parachute, a barometer, a thermometer, and an on-board camera. This thirty-five-pound rocket tilted immediately after leaving the tower, then dove for earth. The engine's roar was punctuated by the explosion of the gas tank when the rocket hit the ground. After Goddard and his team celebrated their flight, posed for photos, and looked for missing parts, they were greeted by a procession of cars led by two ambulances. The noise led many people in the area to conclude that an airplane had crashed. Before Goddard could do anything, the local papers were reporting the "terrific explosion" of Goddard's "moon rocket." Goddard was forced to reveal that he was testing liquid-fueled rockets, and the news spread around the world.

Among the interested readers of this news was Charles A. Lindbergh, then at the height of his considerable fame two years after his historic solo nonstop flight from New York to Paris. Lindbergh was visiting his friend and patron, millionaire Harold Guggenheim, when the news of Goddard's flight came out. Both were anxious to advance aviation, and the chance of flying into space got their attention. When Lindbergh visited Goddard in November, the two spent hours discussing aviation and space travel, and became lifelong friends. Lindbergh began to look for new sources of funding for Goddard, and after obtaining no help from the DuPont Chemical

Company and just a small grant from the Carnegie Institution, Lindbergh called Guggenheim's father, Daniel, who agreed in 1930 to support Goddard's research with the then generous sum of $25,000 a year for two years and possibly four.

Even before the 1929 flight, Goddard wanted to relocate his static and flight tests to a more isolated location, possibly in the western United States, but funding was a problem. Following the flight, the Massachusetts state fire marshal prohibited further rocket tests in the state. For the next few months, Goddard was able to skirt this prohibition by testing rockets on federal land at the Hell Pond area of Camp Devens, an army post near Worcester. With the Guggenheim money, Goddard took a leave from his academic duties and in the summer of 1930 moved his team to Roswell, New Mexico, an area recommended by Lindbergh and others as having the right weather, terrain, and isolation for his work.

The Goddards rented a fifteen-acre property known as Mescalero Ranch two miles northeast of the center of town, where they set up their home and a workshop to assemble rockets. Ten miles out of Roswell on a cattle ranch they found a small depression known as Eden Valley where they set up a launch tower. Just before the end of the year, Goddard and his team of four assistants launched a rocket that flew two thousand feet high, reached a speed of five hundred miles per hour, and returned relatively undamaged thanks to a parachute. By then, the Goddards called all their rockets "Nell," the name coming from a well-known character in a Broadway melodrama of the time. But the Nells of 1931 and 1932 did not fare as well as the first New Mexico Nell, even as Goddard tried to improve his rockets with gyroscopes and exhaust vanes to keep them flying straight.

Daniel Guggenheim died just after the Goddards moved to New Mexico, and estate problems combined with the impact of the Depression depleted the Guggenheim funds. As well, Goddard's patron Lindbergh was occupied with the kidnapping and murder of his infant son. As a result, the Guggenheim funding ended after two years, and Goddard returned to Clark University and his academic work for two years. During this time, Goddard continued his work on pumps and gyroscopes.

Luckily, Harry Guggenheim restored his family's fortunes, and in 1934 he resumed funding to Goddard, although at the reduced rate of $18,000 a year. The grant was only for one year, but it was renewed each year, and the

6. Standing in front of Dr. Goddard's rocket in the launch tower near Roswell, New Mexico, on September 23, 1935, are (left to right): Albert Kisk, Goddard's brother-in-law and machinist; Harry F. Guggenheim; Dr. Goddard; Col. Charles A. Lindbergh; and N. T. Ljungquist, machinist. Charles Lindbergh, an advocate for Goddard and his research, helped secure a grant from the Daniel and Florence Guggenheim Foundation in 1930. (NASA photo)

Goddard group remained in Roswell until 1942. In 1934 and 1935, Goddard and his team launched fourteen A-series Nells. Some failed while others went up as high as a mile. In the following months, he tested K-series rocket motors in static tests, began launching L-series Nells starting in 1936, and flew these rockets and P-series rockets until 1941. An L-series Nell reached eight thousand feet in altitude, the greatest for any Goddard rocket. He tinkered with guidance systems, then pumps, and then tried to deal with problems with combustion chambers in these rockets, which grew to eighteen feet in length. Rocketry required large teams of experts, and a single scientist leading a small team of three or four machinists—especially one like Goddard who changed his mind about what problem was most urgent— couldn't keep pace with the various problems. Clary notes that Goddard also lost time trying to build gyroscopic guidance systems for his rockets, even though they weren't needed in rockets, like the Nells, that were designed to probe earth's atmosphere.

Yet Goddard's firsts included designing and flying the first liquid-fuel rocket engines and rockets with inertial guidance, using a variety of devices to effectively steer rockets, and flying a powered vehicle faster than the speed of sound, among other advances. Though he was loath to share the technical background of much of his work, in 1936 the Smithsonian published an up-to-date summary Goddard wrote of his work called "Liquid-propellant Rocket Development." This paper contained his first public acknowledgement that he had flown the first-ever liquid-fueled rocket ten years earlier.

War Clouds Gather

In the late 1930s, Goddard's work slowed down, in part because of his age and declining health, aggravated by his cigar habit. The coming war began to cast its shadow on Goddard's world as well. In the late summer of 1938, Goddard and his wife took a vacation in Europe but didn't contact any rocket enthusiasts there. On the way home, Goddard stopped in New York and met Harold Guggenheim and officials from the National Advisory Committee for Aeronautics (NACA) and the Guggenheim Aeronautical Laboratory, California Institute of Technology (GALCIT), to talk about rocket research for national defense purposes. A few years later, GALCIT converted itself into the Jet Propulsion Laboratory (JPL), and both JPL and the NACA later became constituent parts of the National Aeronautics and Space Administration, NASA. Under some pressure from Guggenheim, Goddard agreed to work with GALCIT and its leader, Theodore von Kármán, but Goddard backed out soon after he returned home to Roswell. Goddard had already established a testy relationship with GALCIT when one of its leading lights, Frank Malina, then a Caltech graduate student, came to Roswell and visited Goddard in 1936. Although the visit was friendly on both sides, Goddard limited the information he shared with Malina, and later wrote to Caltech scientist Robert Millikan that "I naturally can't turn over the results of my years of investigation, still incomplete, for use as a student's thesis." This insult turned Malina's visit into a legendary incident that was used by both defenders of Goddard and Malina to explain the estrangement between Goddard and GALCIT.

In 1940, Goddard began contacting both the U.S. Navy and the Army Air Corps, the predecessor of the U.S. Air Force, about using rockets should the United States enter the war that broke over Europe in 1939. By the time of the Japanese attack on Pearl Harbor on December 7, 1941, which brought America into the war in both Europe and the Pacific, Goddard was already engaged in defense work for the navy and to a lesser extent the Army Air

Force, which succeeded the Army Air Corps earlier that year. His last attempt to launch a Nell occurred in October 1941, and in 1942 he and his group left Roswell to work for the navy in Annapolis, Maryland.

There Goddard worked on liquid-fueled rocket-assisted takeoff technology for aircraft for the navy and the Army Air Force. His work on what became known as "jet-assisted takeoff units" proved to be of limited use to the armed forces because units that used solid fuels proved to be safer and easier to handle. Moreover, rocket research was moving toward different fuels than the liquid oxygen and gasoline combination that Goddard was used to. As the war went on, Goddard's team turned toward variable-thrust rocket motors. Much of Goddard's time was taken up in frustrating dealings with military bureaucracy, and his own research was also slowed by his declining health and his determination to stick to his own research path. In 1945, he was able to inspect recovered parts from German A-4 rockets, which had greatly improved on his inventions.

At war's end, the Goddard team was transferring its work from the navy to the Curtiss-Wright Corporation when Goddard died of throat cancer in Annapolis on August 10, 1945. Goddard's team continued their work without him at Curtiss-Wright, culminating in the XLR-25 CW-1 variable-thrust rocket motor, which was used in the Bell X-2 research aircraft. This twin-chambered engine marked important advances in rocket engines, but its complexities slowed the troubled X-2 program. Between 1952 and 1956, the X-2 set speed and altitude records that were only broken in the twentieth century by the legendary X-15 rocket plane.

Esther Goddard, with support from Lindbergh, Harold Guggenheim, and Pendray, continued to work hard establishing Goddard's primacy in the field of astronautics. Part of this effort involved publishing both an authorized biography of Goddard upon which much of his legend is still based and a carefully edited set of Goddard's papers. At the center of her work, however, were Goddard's patents. Goddard had been granted 48 patents during his lifetime, and another 35 that he had applied for were later granted. Esther Goddard applied for and received 131 additional patents based on Goddard's work, and so Goddard was granted a total of 214 patents. In 1960, the U.S. armed forces and NASA settled with Esther Goddard and the Daniel and Florence Guggenheim Foundation, which had supported Goddard's research, for $1 million for the use of the technology contained in

Goddard's rocket patents. Perhaps more important than the money in Esther Goddard's eyes was establishing Goddard's position as the inventor of modern rockets, at least in the United States.

Perhaps the biggest lesson that can be drawn from Goddard's rocket work is that rockets are so complicated and have such narrow margins for error that they require large and well-funded teams of designers, engineers, workers, and perhaps most importantly, managers, to succeed. Goddard's insistence on working essentially alone meant that his rockets never rose above eight thousand feet. Those who followed Goddard learned the hard way that even when money, talent, and people are put into rockets, they still can fail. Despite Goddard's failures, his record of success is remarkable, especially given the relatively small resources he worked with. One of the big secrets of his success was that he was able to put failed tests into perspective and draw lessons from each setback.

Goddard's practical accomplishments as a rocket builder and the competing claims for spaceflight concepts have obscured the New Englander's importance as a theorist of spaceflight. Though others may have come up with ideas such as liquid-fueled rockets and staged rockets for reaching space before Goddard did, he came to these ideas independently. He also wrote and talked privately about other ideas such as exploring the moon and Mars with cameras mounted on spacecraft, and also of flying humans in space. He wrote about using planetary atmospheres to brake spacecraft on entry, and of using solar power, ion propulsion, and electrical propulsion for spacecraft. Goddard is remembered today for building and launching the first liquid-fueled rocket in 1926. But his most important contribution to space exploration came in 1920 when he became the first person to successfully raise the idea to a large number of people of using rockets to go into space. The publicity surrounding his important assertion inspired scores of people in America, Europe, and Russia to begin work on their own rockets. Even if Goddard had not gone on to build his own rockets, he would still be an important figure because the publicity surrounding his 1919 paper on rockets inspired the space exploits of the twentieth century.

The American Rocket Society

Goddard's fundamental ideas and work gained importance because they were publicized and awoke the world to the idea that space travel was possible with the rocket. As Pendray later wrote, the publicity about a moon rocket following Goddard's 1919 paper "started a sort of psychological chain reaction all over the world." He inspired other people, and many of them formed rocket clubs. Interestingly enough, the first such group in the United States was probably inspired more by the speculative fiction that also fired Goddard's imagination than by Goddard's work itself. In his history of early rocket societies, Winter writes: "Science fiction was the real parent of the American Rocket Society." A dozen people (including one woman) formed what they called the American Interplanetary Society in a meeting in a New York City brownstone on April 4, 1930. Nine of them were contributors to Hugo Gernsback's pulp magazine *Science Wonder Stories*. David Lasser, the managing editor of the magazine, was elected as the first president, and Pendray, a newspaper reporter who also wrote stories for the magazine, became vice president.

In the early days, the society promoted space travel through its own publication and through public meetings. One of the first things the society did was invite Goddard to give a talk. Although Goddard joined the society, he declined to take an active part in its work. In January 1931, a public meeting that was to be addressed by visiting French aeronautical and space pioneer Robert Esnault-Pelterie drew nearly two thousand people. When the Frenchman had to drop out at the last minute due to illness, Pendray delivered Esnault-Pelterie's speech instead. Because Pendray sported a goatee, most of the spectators never knew that the main speaker never showed up. A few months later, Lasser wrote a book, *The Conquest of Space*, the first book in English on spaceflight.

Inspired by Pendray and his wife Leatrice's visit to meet German rocket enthusiasts and see their rocket test stand in Europe, the society turned to experimentation, even though few of the members at the time had an engineering background. Their first liquid-fuel rocket engine worked successfully in a test stand in November 1932. Pendray, who had become president when Lasser resigned to become a trade union organizer, stepped down in turn from the presidency to concentrate on building the group's second rocket, which flew successfully in May 1933. A year later, the group changed its name to the American Rocket Society and continued testing rocket motors in test stands, holding meetings, and publishing news about rockets and space exploration.

As the decade wore on, the society's membership increased, and more of them were scientists and engineers. Other rocket groups formed around the United States, notably the Cleveland Rocket Society and a group at Yale University, and some were eventually absorbed by the American Rocket Society. Engine testing continued, and on one day in 1938, American Rocket Society members tested three different engines containing technical innovations. One of the society's rockets was exhibited at the 1939 World's Fair in New York, and in 1941 some society members formed Reaction Motors Inc., which won contracts to build rockets during the war and after. During the war, the society republished Goddard's classic 1919 paper, and toward the end of his life, Goddard became involved in the society. He formed a friendship with Pendray, who subsequently worked with Esther Goddard to burnish her husband's reputation. Pendray, for his part, went on to have a successful career in public relations—he is credited with inventing the idea of the time capsule—and wrote a great deal about the early days of rocket research in the United States.

"The American Rocket Society . . . was born in that wave of Goddard-engendered enthusiasm; but the beginnings of its experimental program did not stem directly from Goddard's work," Pendray wrote. "These [rocket society] efforts brought forth a group of men with experience and know-how who were ready and willing to take leadership positions in the modern rocket and missile age. And perhaps equally important, the early rocket experiments helped to promote an ever-mounting pitch of interest and enthusiasm."

4. Hermann Oberth and Early German Rocketry

There is no doubt: The moment is here, the hour has come, in which we may dare to undertake the attack on the stars with real prospects of results. It is clear that the armor of the earth's gravity will not lightly be pierced, and it is expected that it will cost to break through it, much sacrifice of time, money, and perhaps also human life.

German spaceflight pioneer Max Valier

The dreams of aviation and space travel weren't confined to America and Russia early in the twentieth century. Although the Americans Orville and Wilbur Wright were the first to fly a heavier-than-air powered craft, they weren't far ahead of many people in Europe. Indeed, many of the advances in aviation between the Wrights' historic 1903 flights and World War I originated in Europe, especially France. One of those French pioneers of flight was Robert Esnault-Pelterie, who was barely twenty-two years of age when the age of aviation opened in 1903.

Esnault-Pelterie, also known as "REP" after his initials, earned a degree in science at the Sorbonne in Paris in 1902, the same year he was granted his first patent, for an electrical relay. Esnault-Pelterie's interest in aviation dates to April 1903, when he attended a lecture where aviation pioneer Octave Chanute discussed his own work and the gliders being tested by the Wrights as they worked toward their historic first powered flight in December of that same year. "Having started in aviation as early as 1903," he recalled, "in an attempt to check the Wright brothers' [glider] results, I promptly abandoned their biplane type to devise the first monoplane, bearing in front a seven radial cylinder engine; at the rear were two rudders and a fin."

REP went on to build several aircraft that incorporated innovations to aircraft design that endure to this day. He built the first all-metal monoplane, designed the first radial aircraft engine, and originated the control stick, used ever since in aircraft, in addition to making advances in landing gear. Before the onset of war in 1914, he also started an aircraft manufacturing company, though this venture didn't succeed. By then, Esnault-Pelterie was thinking beyond aircraft.

"When flying became a fact," he wrote later, "having once been only a dream, it was apparent to me, as one who remembered the time when there were even no automobiles, that it would develop rapidly, and I wondered what the next stage might be. Once the atmosphere had been conquered, there remained nothing more but to strike out into the empty space of the universe." Such thoughts came naturally to Esnault-Pelterie, who was a born tinkerer with a wide interest in many facets of science and engineering. Born in Paris on November 8, 1881, eleven months before Goddard, Esnault-Pelterie was known as a sportsman and outdoors enthusiast who loved to spend time in his machine shop building parts for aircraft and other equipment. For example, he loved to camp in a retreat in the Pyrenees Mountains, and his "tent" included a raised floor, heat lights, running water, a bathtub, a telephone, and a radio that was connected to a hidden microphone (useful for playing jokes on guests).

In 1912, Esnault-Pelterie gave lectures in St. Petersburg, Russia, and in Paris in which he speculated on space travel. In his talks he emphasized the importance of changing mass ratios in rockets and the fact that rockets can travel faster and farther as their fuel is burned and can operate in a vacuum. He also spoke about trajectories to the moon, Mars, and Venus, and said atomic energy could open the door to interstellar travel. This lecture was based on his speculations on space travel, which he began to pursue about 1908. Tsiolkovsky and Goddard were also thinking about these concepts, and Dr. André Bing of Belgium had received a patent in his home country for an apparatus that included staged rockets. Although a truncated version of Esnault-Pelterie's talk was published in France, it did not stir the public interest that Goddard's 1919 publication did, perhaps because of its obscure title or because it was not backed up by an institution like the Smithsonian. REP put aside his thoughts about space travel during World War I, but by the late 1920s he was lecturing on space exploration again and working on

7. French postage stamp honoring Robert Esnault-Pelterie.

his master work, *L'Astronautique* (Astronautics). This work, issued in 1930 with a supplement five years later, covered a wide variety of topics in the new science of astronautics, including rocket structure and motion and spacecraft design and trajectories. In the book's title Esnault-Pelterie coined the word *astronautics*. Though REP's work entitles him to stand alongside other fathers of both aviation and space travel, his contributions to these fields are unfortunately little known outside his home country.

In the 1930s, Esnault-Pelterie experimented with different liquid fuels for rockets, and one test in 1931 resulted in an explosion that cost him four fingers. His research in aviation and astronautics encompassed many areas, including metallurgy, properties of liquids, and electricity, and he earned more than two hundred patents in his career. Much of his research on missiles during the 1930s was supported by the French Air Ministry, but it was classified and then lost during World War II. The war ended Esnault-Pelterie's research, and he retired to obscurity in Switzerland. Disputes with the tax department made his retirement difficult, but he did live to see the launch of the first earth satellites before his death on December 6, 1957.

During his life, Esnault-Pelterie also tried to popularize space travel. Together with his friend, the banker André Hirsch, he established the REP-Hirsch International Astronautics Prize in 1928, an annual award by the French Astronomical Society for the best original scientific work advancing the science of astronautics. Esnault-Pelterie encouraged Goddard to seek the prize, but Goddard declined, probably because it would have required him to supply paperwork to back up claims to the prize. In 1936, the final REP-Hirsch Prize was awarded to the American Rocket Society for its experimental work. Although the prize was awarded for only a few years, its existence helped stir interest in the new science, and the first prize in 1928 encouraged one of the greatest pioneers of space exploration.

Hermann Oberth

The winner of that first REP-Hirsch Prize was a schoolteacher from Transylvania named Hermann Julius Oberth. He was honored for work that had inspired space travel enthusiasts and rocketeers in Germany and around Europe. For years, Oberth had thought about flight into space, and in 1923, he published his best-known work, *Die Rakete zu den Planetenräumen* (By Rocket to Interplanetary Space), which became a bestseller among European space enthusiasts. "It is reassuring to see that science and progress suffice to overcome national prejudice," Oberth said in accepting the REP-Hirsch Prize. "I can think of no better way to thank the Société Astronomique de France than to pledge myself to work on behalf of science and progress and to judge people only on their personal merits." The turbulence that afflicted Europe in the first half of the twentieth century had touched Oberth like many others, and it would later dash the hopes he expressed in 1928.

Hermann Oberth was born on June 25, 1894, in Sibiu, a small town in the Transylvanian Alps, which were then part of the Austro-Hungarian empire. The area had been settled seven centuries before by a group of Saxon Germans who still lived as Germans and called the town Hermannstadt. Two years after Oberth's birth, when his father, a doctor, got a job as medical director of a hospital, the family moved to the nearby town of Sighisoaro, known in German as Schässburg.

The young Oberth grew up curious about the world around him. The gift of a small telescope allowed him to follow trains passing through the area, but he was soon using it to look at the moon and other objects in the sky. One day, his mother gave the eleven-year-old boy her copy of Jules Verne's *From the Earth to the Moon*, which filled his mind with visions of traveling to the body he had admired so many times through his small telescope. "I was fascinated by the idea of space flight, and even more so, because I succeeded in verifying the magnitude of the escape velocity," he later wrote. He was struck by Verne's recounting of Newton's laws of motion, and he set out to prove to himself that every action had an equal and opposite reaction, by pushing off on a nearby bank. As other experiments followed, Oberth began to dream of flying to the moon. But his dreams ran into the reality of his father's expectations: he was set on his oldest son following him into the medical profession.

In the fall of 1913, Oberth entered the University of Munich as a medical student. Luckily, he was also able to sign up for courses in astronomy and physics. The following summer, his plans were disrupted along with those of millions of others when Europe was plunged into World War I. Oberth returned home to Schässburg and joined the army of Austria-Hungary, which was allied with Germany, starting out as a soldier in the infantry. Everyone believed that the war would be short, but it dragged into months, and when Oberth himself was injured in an accident he was transferred to a field ambulance unit, where he gained experience in treating war casualties. As Oberth said later: "This was a piece of good luck, for it was here that I found that I should probably not have made a good doctor." His wartime luck extended to his personal life when in 1918 he married Mathilde Hummel, another Transylvanian German from Schässburg.

Between flurries of activity for his ambulance unit, Oberth had time to pursue his interests in space and rockets. Using sedative drugs and small water tanks to simulate weightlessness, Oberth tried to anticipate the physiological effects of flying in space. He also spent time designing a rocket that would end the war on a favorable basis for Germany and Austria-Hungary. He brought designs for his rocket, eighty-two feet high and designed to carry explosives a long distance, to the attention of German authorities, but they rejected it. Meanwhile, the war took many turns. Romania, which had territorial designs on Transylvania, entered the war against Germany and Austria-Hungary. German spirits soared when the Russian tsar fell in the first Russian Revolution of 1917, and the Bolshevik government installed during the second revolution that year sued for a favorable peace with Germany. But the same year, the United States entered the war on the side of Britain and France. After German offenses in the spring and summer of 1918 failed, German forces fell before the newly strengthened allies in the west, and by November, the Austro-Hungarian empire had fallen apart and Germany surrendered. In the confusion and bargaining that followed the armistice, Oberth was discharged from his military duties, but Transylvania had been ceded to Romania, which imposed its language on the area.

Oberth had by then finally convinced his father to allow him to pursue his interest in space exploration, and he returned to Munich to resume his studies. But his new status, holding a Romanian passport in a Germany troubled by its humiliating military defeat, political tumult, and runaway inflation, meant that Oberth had to argue to even be admitted into Germany. Then he had difficulty finding housing for himself, his wife, and his children, and his work on rockets and space travel was not well accepted by his professors. He moved from Munich to Göttingen University and finally in the fall of 1921 to the University of Heidelberg, by which time he was designing a two-stage rocket. In 1922, Oberth submitted his thesis on rockets and space travel to the university, but it refused to grant him the doctoral degree he sought.

At the same time, Oberth read newspaper accounts of Robert Goddard's work, so that year he wrote a letter to Goddard saying that he had been working for years on rockets and would like to get information on Goddard's work. Goddard sent him a copy of his famous 1919 paper on rockets in return. "I think that only by common work of the scholars of all nations can be solved this great problem," Oberth wrote in his stilted English. Goddard's letter of reply is lost.

The astronomer Max Wolf suggested that Oberth publish his thesis, so Oberth added some popular material describing his ideas on space exploration to his mathematics-filled thesis and began submitting his work, *Die Rakete zu den Planetenräumen* (By Rocket to Interplanetary Space), to German book publishers. After many rejections, the Munich publisher Rudolf Oldenbourg agreed to publish the ninety-two-page book if Oberth paid the printing costs. The first part of the book contained technical information about rockets and space travel, and the second part concerned the design of a two-stage liquid-fueled rocket for exploring the upper atmosphere. The third section described a spacecraft, the Model E, that could carry humans to other worlds. Such spacecraft could be put in orbit around the earth and used to transmit communications signals, observe earth, and carry large mirrors to reflect light onto the earth at night to light towns and regulate weather. Oberth's book also included an appendix describing Goddard's work in the United States.

While his manuscript was making the rounds of Germany's publishing houses, Oberth had studied Romanian, earned his teaching certificate, and returned to his hometown of Schässburg before landing another teaching job in the nearby town of Mediasch. When the book was published in 1923, the first printing sold out quickly, but academic critics attacked Oberth's claims. Some of the attacks were similar to those leveled at Goddard: rockets can only work if they have air to push against. "We believe that the time has not come for delving into such problems as these and probably never will come," a distinguished reviewer wrote in an engineering journal. In reply Oberth said that many of his academic critics were simply too busy performing their research and teaching duties to give his ideas the consideration they required. Ironically, one of Oberth's critics was Robert Goddard, who was convinced the schoolteacher had stolen his ideas. Word came to Oberth, who in later years denied any influence from the American. A German biographer of Oberth even later claimed that Goddard's work had been inspired by Oberth. Lost in the accusations was the fact that many of the ideas contained in Oberth's book went far beyond Goddard's public suggestion of sending flash powder to the moon. With his suggestion that rockets provided a feasible way of sending humans into space, Oberth fanned the interest in space travel first raised by Goddard with his modest moon rocket, and raised it to new levels.

Rocket Fever Grips Europe

Despite the criticism, Oberth's book went into additional printings and awoke enthusiasm for space travel around Europe. The foundation for this enthusiasm had been laid in part by the fiction of Verne, Wells, and others, as well as by the search for new ideas in a Europe hungry for diversion after the hard years of World War I. Oberth also built on other people's speculations on space travel. In Germany, an eccentric inventor named Herman Ganswindt had proposed a spacecraft in 1891 that was powered by dynamite cartridges exploding against its walls, and his idea had received a great deal of publicity. In the 1920s others were inspired to write about space travel. A German architect named Walter Hohmann published a book, *Die Erreichbarkeit der Himmeslskörper* (The Attainability of the Celestial Bodies), in 1925 that was even more technical than Oberth's, examining issues related to a rocket's orbits and trajectories in space. The Hohmann Transfer, a still widely used means of moving from one orbit to another to reach another body in space, bears the name of this space pioneer. An Austrian military engineer, Hermann Potocnik, using the pseudonym Hermann Noordung, wrote in great technical detail about his ideas for a wheeled space station in his book, *Das Problem der Befarung des Weltraums* (The Problem of Spaceflight), which came out in 1929. An aerodynamicist at the Vienna Polytechnic Institute named Eugen Sänger began work on a hybrid aircraft and rocket that could fly high in the atmosphere at great speeds and also conducted research on rocket engines. Sänger continued to work on the vehicle, which he named Silverbird, with the assistance of mathematician Irene Bredt, who also became his wife. Sänger and Bredt's concept became one of the most alluring and elusive ideas in the early history of space exploration.

Perhaps most importantly, Oberth's work inspired people who were even more capable of disseminating his ideas. One was Max Valier, an Austrian writer living in Munich who had established himself as a popularizer of science. Valier began to write and speak widely about Oberth's work, and his best-selling 1924 book, *Der Vorstoss in der Weltenraum* (A Dash into Space), made Oberth's ideas accessible to a wider audience. One person whom Valier got to know in Munich about this time was a former soldier and struggling politician named Adolf Hitler, who later dismissed Valier as a "dreamer." A twenty-year-old German writer named Willy Ley was inspired by Oberth but put off by many features of Valier's book, including a number of errors and technical sections in daunting fine print. "With the enthusiasm peculiar to that age, I sat down and wrote a small and formula-free book on the same subject," Ley later recalled. His 1926 book, *Die Fahrt in der Weltenraum* (The Flight into Space), was also a success. Nevertheless, it was Valier's skill at gaining publicity that proved crucial to building the spaceflight movement in Germany and across Europe in the 1920s.

Together with the publicity that surrounded Goddard in 1920, the work of Oberth, Valier, and Ley popularized the idea of the rocket as the vehicle for piercing outer space. In his recent survey of science fiction writings and movies, historian Frank Winter found that rockets were the prime mode of transport into space depicted in these speculative works after 1920, a fact he attributes to the efforts of Goddard and the Germans: "There is no question Oberth's book, with its detailed plans of a manned, liquid-propellant staged rocket contrasted with Goddard's unmanned solid-propellant stage rocket [of 1920] placed the state of the art of astronautics literally on a whole new plane with far wider possibilities offered than by Goddard."

The work of Oberth and Valier inspired the formation of small groups of space enthusiasts in Germany and Austria, and in 1927 Valier and Ley began working with others to set up their own group dedicated to promoting the exploration of space. On July 5, 1927, nine men and one woman gathered in the parlor of the Golden Scepter tavern in the industrial town of Breslau for the first meeting of the Verein für Raumschiffahrt, also known as the "VfR" or German Rocket Society. Although Ley is now considered a founder of the group, he missed this first meeting. Valier declined the presidency of the group, and instead a Breslau engineer and church administrator named Johannes Winkler took the job. Indeed, the meeting was held in Breslau, now the Polish town of Wroclaw, because Winkler volunteered to handle the process of registering the group with authorities. When Winkler took the paperwork to the Breslau court, his application was refused because the word *Raumschiffahrt*, which translates as space travel, did not then exist in the German language. Once the group's charter was amended to include a definition of the word, the VfR was incorporated.

From this tiny core group, VfR rapidly expanded within two years to a thousand members from around Germany, including Oberth and Hohmann, as well as others from outside Germany. Many joined to receive the society's journal, *Die Rakete*, but others wanted to begin the practical work of making space travel a reality. Characteristically, the first one out of the gate was Max Valier. In a move that Ley termed "colossal nonsense," Valier contacted Fritz von Opel, who had won fame and fortune by building a line of inexpensive cars. Valier persuaded von Opel, and out of their conversations grew an effort to build the world's first rocket-powered cars. Both Valier and Opel were out to gain publicity, and both wanted to proceed as quickly as possible. Valier contacted Friedrich Sander, who manufactured solid rockets used as navigational aids or for throwing lines during coastal rescues. The first test of a rocket car took place on Opel's racetrack on March 15, 1928. Opel driver Kurt Volckhart took the wheel of a car propelled by two rockets. His drive was relatively slow, only 450 feet in thirty-five seconds, and neither he nor Opel were impressed. They almost called off further testing when Sander demonstrated the power of one of his rockets when flying free, and a second drive of the car with more rockets attached was sufficiently impressive to persuade Opel to order a specially built rocket car.

The new car, known as the Opel-Rak 1, was a converted racing car equipped with Sander rockets instead of an internal combustion engine. On April 12, the car made successful runs powered by six and then eight rockets, reaching speeds of 55 miles per hour over distances up to three thousand feet. A few weeks later, Opel himself took the wheel of a more streamlined Opel-Rak 2, equipped with twenty-four rockets, which propelled him to a speed of 125 miles per hour. In the blaze of publicity that accompanied this feat, Opel promised more ambitious vehicles. Opel-Raks 3 and 4 were rocket-powered railway cars that both derailed in their runs, and railroad officials refused to lend their rails for the proposed Opel-Rak 5. Opel, who by then was working independently of Valier, flew a rocket-powered aircraft in September 1928 but abandoned this avenue after his one successful flight. One reason may have been that a group of German glider enthusiasts had flown the first rocket-powered aircraft three months earlier. "What had started as a kind of scientific movement was almost smothered under a series of publicity stunts," Ley wrote later. These "stunts" drew attention away from work on liquid-fueled rockets that could actually advance the art of rocketry beyond where solid rockets seemed stalled. Yet the publicity stunts were far from over.

This time the man at the center of the publicity campaign was Hermann Oberth. In the fall of 1928, the high school teacher had been called from Transylvania to Berlin by none other than Fritz Lang, the greatest German film director of the time, who had decided to make a film about space travel. "It is almost impossible to convey what magic that name had in Germany at that time," Ley wrote of Lang. "It is not an exaggeration to say that a sudden collapse of the theater building during a Fritz Lang premiere would have deprived Germany of much of its intellectual leadership at one blow." Lang's *Metropolis* is still considered one of the great classics of film, and so when he decided to shoot a film based on a script by his wife Thea von Harbou titled *Frau im Mond*, or "Woman in the Moon," the hopes of space enthusiasts rose. Lang hired Oberth as a technical consultant for the film. Oberth, for his part, had written an upgraded version of his first book, which came out in 1929 as *Wege zur Raumschiffahrt* (Road to Space Travel). For it and his other contributions, Oberth won the first REP-Hirsch Prize, along with prize money.

Oberth and others hoped that the Lang film would free up money for rocket research, and Oberth began discussions with Lang and the Ufa Film Company about building a liquid-fueled rocket whose launch at the time of the film's premiere would further publicize *Frau im Mond*. Ufa didn't want to spend money on a rocket, so Lang put some of his money together with

Oberth's REP-Hirsch prize money to finance the rocket. This project was ill starred for a number of reasons. Though Ley believed that Oberth was the greatest rocket theorist of the day, "he lacked the ability of an experienced engineer to say in advance what can be done and what probably cannot be done." As well, Oberth was unused to the hustle and bustle of Berlin in 1929. Instead of working through an engineering group to hire assistants, Oberth placed an advertisement in the paper, hiring two men who answered it. One was Rudolf Nebel, who claimed to be a fully qualified engineer who as an air ace in World War I had experimented with rockets in aircraft, qualifications that were exaggerated. The second was Alexander Shershevsky, a Russian flying student who had decided not to return to the Soviet Union after overstaying his authorization to travel. As Ley put it, "The trio, consisting of a slightly bewildered theorist, a professed militarist, and a Bolshevist accidentally in disgrace, worked together, or tried to."

Given the trio's limitations, and the fact that only one person, Robert Goddard, had actually succeeded in building a liquid-fueled rocket up to that time, Oberth's effort to construct a rocket in time for the movie's premiere was not likely to succeed. While Oberth ran experiments on bringing liquid oxygen and fuel together for combustion, he made progress, but one experiment ended in an explosion that nearly cost him the sight in one eye. After a short convalescence, Oberth and his group pressed on, but their troubles increased. Authorities refused permission to launch the rocket at a seashore launching site. Finally, when Oberth realized that only a few weeks remained before the premiere, he tried and failed to build a simpler rocket. When *Frau im Mond* opened on October 15, 1929, there was no rocket launch. Oberth, his "nerves almost shattered" by the explosion, quickly left Berlin for his home in Mediasch. Except for a short-lived return to Berlin in 1930, Oberth spent the next few years teaching "practical engineering" such as locksmithing at the local high school and relegating rocket research to the side.

As for the movie, its popularity suffered from the fact that it was a silent film at a time when "talkies" were taking the world by storm. But Lang, and *Frau im Mond*, nevertheless left their mark on spaceflight. Just before the rocket's realistic-looking launch in a hail of sparks and smoke, the words "10 seconds to go" appeared on the screen, followed by "6-5-4-3-2-1-0-FIRE."

Fritz Lang had invented the countdown, which German rocketeers picked up for their launches.

Nine days after the film's premiere, stock prices tumbled across the Atlantic in the New York Stock Exchange, and soon the world was feeling the effects of a worldwide depression that displaced the prosperity of the 1920s. Along with so much else, the hopes of space exploration advocates suffered. Winkler suspended publication of *Die Rakete* and stepped down as president of the VfR, Oberth succeeding him. Recovering the equipment that Oberth had used for his rocket and for the film, Oberth and other society members began to experiment with liquid-fueled engines, starting with a small rocket engine that Oberth called the Kegeldüse, based on the German word for the shape of the engine: a cone. Oberth had had several of these engines built while he was working on the rocket for the film premiere, and VfR members pressed these engines into service. Among these VfR members were Nebel, who had decided to stick with rocket research despite losing his paying job with Oberth, a young engineer named Klaus Riedel, and an even younger engineering student named Wernher von Braun. After Oberth returned home to Mediasch, a Kegeldüse was test-fired successfully in 1930 for German scientists and media. As a result, the VfR drew crowds when it held its first public meeting in Berlin, and Max Valier turned from solid- to liquid-fuel rockets. Valier had meanwhile built a new rocket car with a liquid-fuel rocket, but in the first test in April it moved slowly due to poor combustion in its engine. On May 17, Valier was running the engine in a test stand when it exploded, killing him. "He died while engaged in his first really useful experiment," Ley wrote, "although the idea of mounting his motor in a car was, of course, ridiculous."

Raketenflugplatz

Though Valier's death spurred some people to call for a ban on rocket testing, the VfR members pressed on with work on a liquid rocket of their own, known as "Mirak," short for "Minimumrakete." The first Mirak exploded in the fall of 1930. Several VfR fund-raising efforts failed until two wealthy members decided to help underwrite the research work. Armed with this money, Nebel and Riedel of the VfR went looking for a place to do their experimental work. They found an unused piece of land on the site of a

8. A group of rocket experimenters in Germany in 1930. From left to right are Rudolf Nebel, Dr. Ritter, Mr. Baermueller, Kurt Heinish, Hermann Oberth, Klaus Riedel, Wernher von Braun, and an unidentified person. (NASA photo)

World War I army garrison in the working-class suburb of Reinickendorf, on Berlin's northern fringes. The city of Berlin rented the three-hundred-acre tract, which also conveniently had protected bunkers, to the VfR in September 1930, and Nebel named it Raketenflugplatz, or rocket aerodrome. Nebel and Riedel moved into a building on the land, sharing it with rockets and spare parts. Work began on a second Mirak, which exploded early in 1931. However, subsequent rockets, named "Repulsors" to distinguish them from solid rockets, began flying in May.

In April of that year, G. Edward Pendray and his wife visited the Raketenflugplatz and were shown a test run of a rocket motor. Duly impressed, they returned to America and got the American Rocket Society into the business of testing rocket engines. In this way, the German rocketeers probably did more than Robert Goddard himself to spur what became America's homegrown rocket talent.

The successful flights of the Repulsors were not the first liquid-fuel flights in Europe, however. Johannes Winkler, who had gone his own way from the VfR in 1929, launched a liquid-fueled rocket on February 21, 1931, at Dessau. Ley later wrote that since Goddard did not formally publish the results of his liquid-fuel rocket work until 1936, "we naturally took it to be the first liquid-fuel rocket anywhere." In his history of the early rocket societies, historian Frank Winter has noted that Friedrich Sander, the builder of the solid rockets used by Opel, apparently launched a liquid rocket of his own on April 10, 1929.

In any case, for a little more than a year, the Raketenflugplatz in Berlin became the center of German rocketry. Even Winkler set up shop there for his own tests. As the Depression took hold in Germany, Nebel, Riedel, and their associates used all their wiles to keep the place going. They took lumber from abandoned rail cars and procured a typewriter from a pawnshop to send appeals for funds and request tax concessions on the gasoline they used for their rockets and cars. Unemployed craftsmen were given shelter in buildings at the site and put to work building rockets and test stands. Nebel obtained funding support through the Arbeitsdienst, the German version of the U.S. Works Progress Administration. Nebel even declared bankruptcy in one scheme to raise money. Potential donors were invited to view launches, and others were charged admission. Nebel pursued anyone he thought could help, and the most prominent name was Albert Einstein. Although Nebel seems not to have impressed the famed physicist, Einstein's son-in-law visited and left this recollection: "The impression you took away with you was the frenzied devotion of Nebel's men to their work. Most of them were [like] officers living under military discipline. Later, I learned that he and his men lived like hermits. Not one of these men was married, none of them smoke or drank. They belonged to a world dominated by one single wholehearted idea."

The beginning of the end of the Raketenflugplatz came in November 1931 when a Repulsor flew nearly five thousand feet high but crashed into a nearby police station after its parachute failed. Though the damage to the station was light, the police banned rocket flights and then permitted them only under strong restrictions. The weather that winter was severe, but the economic and political storms that buffeted Germany culminated in the appointment of Adolf Hitler as chancellor of Germany in January 1933. That winter, as unemployment soared, membership in the VfR dropped off, and wealthy benefactors no longer felt secure enough to support the rocket experiments. Ley himself found it more profitable, both from a financial and promotional viewpoint, to go on the lecture circuit.

Nebel had one more scheme to keep his rockets flying. He met Franz Mengering, a self-styled engineer from the city of Magdeburg who believed fervently in one of the pseudoscientific theories that were making the rounds in the 1930s, the Hollow Earth Doctrine. This theory held that life on earth actually existed inside a hollow sphere, that the universe was an illusion, and that a rocket flight would lead to the antipodes on the other side of the hollow earth sphere. In October 1932, Mengering persuaded leading officials in Magdeburg to pay for a rocket flight to test the Hollow Earth Doctrine. In January 1933, the city signed an agreement with Mengering and Nebel to underwrite a manned rocket flight that June during the religious holiday of Pentecost. Nebel put everyone at Raketenflugplatz to work, but the rockets fell short of the ambitious plans set for June. The full-scale rocket that could carry a man was never built. Nevertheless, a series of rocket tests in what has become known as Project Magdeburg took place near the city that June, all ending in one sort of failure or another. In return for partial fulfillment of his promises, Nebel received only partial payment. Nevertheless, he and his colleagues continued their work testing small rockets that summer in isolated locations near Berlin. Because of the police restrictions, they couldn't fly at the Raketenflugplatz, however, and the next year, Nebel, who maintained that he ran the Raketenflugplatz on his own without affiliation with the VfR, was forced to give up the testing ground when he couldn't pay a 1,600-mark water bill, most of which was due to leaky faucets.

The VfR itself was wound down by Ley and others early in 1934, most of its remaining members joining another group, the Society for Progress in Traffic Technics. A legend has arisen that all writing and talk in public about rockets and space travel stopped in Germany at that time under orders from Hitler's Nazi government. Winter has noted that there is no record of such an order, however, though it is likely that the Nazis prohibited all talk of the military rocket work already under way in Germany. In any case, Ley left Germany for the United States in 1935. Articles about space travel appeared in the traffic technics society's journal until the eve of World War II.

The VfR was not the only organized group of space enthusiasts in Germany in the 1930s. Winkler founded a group in Dessau in 1932 that didn't last long. Another small group of rocket enthusiasts near Hanover produced recruits for the German military rocket program, one of them, Konrad Dannenberg, later playing a major role in the U.S. space program. In 1937, an astronomer named Hans Kaiser began a Society for the Exploration of Space in Breslau, original home of the VfR. Krafft Ehricke, who became a major figure in space exploration, belonged to this group. It expanded to various parts of Germany and published articles until paper shortages during the war put a stop to its work. Several rocket enthusiasts flew rocket mail delivery flights in the early 1930s to raise funds and publicize their work, including Friedrich Schmiedel in Austria and Reinhold Tiling and Gerhard Zuckel in Germany.

One of the last foreign visitors to the Raketenflugplatz before it closed in 1934 was an engineer-contractor from the suburbs of Liverpool named Phillip Cleator. The previous fall, Cleator and a handful of other British space enthusiasts had founded the British Interplanetary Society (BIS). Cleator was, like many other spaceflight enthusiasts, a science fiction aficionado, and shortly after his visit to Germany a story of his had been published in Hugo Gernsback's *Science Wonder Stories*. The BIS began in Liverpool and was run by Cleator during its first four years until its headquarters was moved to London in 1937 and Prof. Archibald M. Low became president. One of the members of the new executive was a young space enthusiast named Arthur C. Clarke.

Those BIS members who wanted to experiment with rockets ran into Britain's Explosives Act of 1875, known as the Guy Fawkes law after the infamous 1605 gunpowder plot. Prohibiting experimentation with gunpowder and other explosives, it was interpreted to include liquid-fueled rockets. Eric Burgess, who had helped start another group in Manchester that later became part of the BIS, publicly experimented with rockets in 1937, and so along with three associates was charged under the act. The trial attracted publicity, and Burgess and his associates were freed on condition that they cease their experiments. This was frustrating to British space enthusiasts because the government had looked the other way when Zuckel came to England in 1934 and launched several rockets carrying mail.

The BIS made its own contribution to space exploration, but not through rocket development. Cleator wrote *Rockets through Space*, which publicized space exploration in the English-speaking world. The new leaders of the BIS in London set to work designing a spaceship to fly to the moon. Early in 1939, the society's journal published Harry E. Ross's groundbreaking article describing the results of the spaceship study. The spaceship, which was based on six stages of solid rockets arranged in "cellules," would stand more than one hundred feet high and carry a crew of two or three to the moon. Although the design was wildly overoptimistic in many ways, it did consider many of the practical problems that would have to be dealt with if a first voyage to the moon were to become a reality. The coming of war that fall forced the BIS to halt its activities, but it resumed its work publicizing the latest thinking about rockets and space travel after the war ended in 1945.

The strict application of the Explosives Act in 1937 may have reflected the war clouds gathering over Europe. Though in France, Esnault-Pelterie was already conducting classified rocket research for the military, the first signs of the coming war had in fact become visible at Berlin's Raketenflugplatz earlier in the decade. In the spring of 1932, three German army officers donned civilian clothing and paid a visit to the Raketenflugplatz. Artillery colonel Karl Becker and his two staff officers, Maj. Wolfram Ritter von Horstig and Capt. Walter Dornberger, came looking for scientific data on the performance of the VfR's rockets. Despite the many tests that were conducted there, the VfR lacked the proper instruments to record engine performance, however, and so the officers left disappointed. Always on the lookout for opportunities to land an army contract for rocket research, Nebel approached them to arrange a demonstration of a rocket, this time at an army range. At the June 1932 demonstration, Becker, Dornberger, and their colleagues were again disappointed when a Repulsor rose only one hundred feet during the flight, which was conducted by Nebel, Riedel, and von Braun. Though the rocket's failure and Nebel's showmanship put off the army officers, they were impressed by von Braun, and by December the young student had left the Raketenflugplatz to continue his work at the army proving ground at Kummersdorf. The young von Braun's move was the first in a major transformation of German rocketry that would change the face of rocketry and space exploration.

5. Von Braun, Dornberger, and World War II

The more technology advances, the more fateful will be its impact on humanity. But if the world's ethical standards and moral laws fail to rise and be adhered to with the advances of our technological revolution, we run the distinct risk that we shall all perish.

Wernher von Braun

On December 1, 1932, the twenty-year-old engineering student Wernher von Braun reported to work for his new job at the German army's Kummersdorf proving ground southwest of Berlin. That simple act signaled the beginning of rocketry's move from the realm of theorists and lone inventors to the large military-industrial-academic teams required to build rockets capable of flying into space. In contrast to von Braun, Konstantin Tsiolkovsky and Hermann Oberth were developing their own ideas of spaceflight without outside help, and Robert Goddard was spending the decade learning the limitations of working alone on rockets. Meanwhile, other rocketeers in the United States, Germany, Russia, and elsewhere were still working in small groups.

By the time von Braun arrived at Kummersdorf, the army had been looking into rockets for three years. Inspired by Oberth's work and the efforts of Max Valier and others to popularize rockets and space travel, Karl Emil Becker, a lieutenant colonel in the army's ordnance office with a doctorate in engineering, decided it was time for the army to see if the rocket, displaced in the nineteenth century by other forms of artillery, offered new potential as a weapon. Advances in solid rocket technology also offered more immediate hope for powerful new artillery weapons to Becker, who like the rest of the German military establishment chafed under the restrictions of the Treaty of Versailles. Rockets were of interest because they weren't specifically prohibited by the treaty, though the German military was also conducting research into aircraft, tanks, and poison gas, which were prohibited by the treaty. Becker had worked on the "Paris Gun" of World War I, which could lob shells eighty miles but had worn out after only 320 rounds. He had helped Rudolf Nebel and the other VfR rocket enthusiasts set up their Raketenflugplatz, but had quickly been put off by Nebel's self-promotion. In 1931, Becker contacted Paul Heylandt, a manufacturer who had backed the rocket-car project that claimed Valier's life, to conduct limited studies on liquid-fueled rocket engines. And in 1932, Becker's group gave the VfR enthusiasts one more chance. By then, Becker had added several other officers to his group, all university trained in engineering. The most important was Walter Dornberger, a pharmacist's son who had fought in heavy artillery units during World War I and had become a spaceflight enthusiast after reading Oberth's work.

Although the army officers felt that the VfR's rocket demonstration fell so far short of Nebel's promises that they chose not to pay him, they were impressed enough to begin building a test stand for liquid-fuel rockets. When Nebel and von Braun met Becker in a futile effort to obtain payment for the test, Becker took note of the young von Braun, whose common sense, aristocratic background, and scientific approach to problems appealed to him. He soon offered von Braun a job with the army.

Von Braun's employment by the army to build rockets was not only significant for being the first marriage of rocketry expertise and the massive resources needed to build large rockets, but because of what von Braun himself brought to the effort. Though the engineering student would not duplicate the theoretical leaps of Tsiolkovsky, Goddard, Oberth, or Esnault-Pelterie, von Braun opened up the way to space by combining the latest knowledge of astronautics with gifts of leadership, dedication, and salesmanship. It is almost impossible to imagine the space age of the twentieth century without Wernher von Braun. It was his unique fate to have to overcome his participation in the century's darkest deed in order to make possible its most sublime achievement.

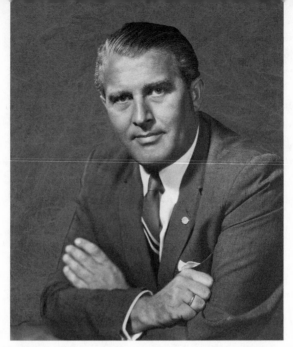

9. A portrait of Dr. Wernher von Braun as director of NASA's Marshall Space Flight Center in 1960. (NASA photo)

Wernher von Braun

Among the pioneers of space travel, von Braun stands out for his aristocratic origins. His father, Baron Magnus von Braun, owned estates in Silesia and East Prussia, was trained in the law and economics, and spent most of his career as a high-level civil servant. When the second of his three sons, Wernher, was born on March 23, 1912, the baron was a provincial councillor in Wirsitz in Posen province, where he and his family lived. Two years earlier the baron had married Emmy von Quistorp, a product of another aristocratic family with Swedish origins. When Germany was forced to yield Posen province to Poland after World War I, the von Brauns moved to Berlin, where the baron continued his government work. By the time Wernher was hired by the army, his father was minister of agriculture in the "cabinet of barons" that presided over the closing days of the Weimar Republic government, which ruled Germany between World War I and Adolf Hitler's rise to power.

Von Braun remembered his childhood as lively and carefree despite the political and economic storms buffeting Germany. The baroness taught young Wernher to play the piano, and he later wrote several short pieces for the instrument and became a skilled cello player. At home and at school, the family practiced speaking multiple languages. Von Braun's first school was the French Gymnasium in Berlin, but when the boy's marks, including those for physics and mathematics, fell short of expectations, his parents transferred him to a boarding school at Ettersburg Castle near Weimar where advanced teaching methods were employed. When about the same time von Braun was confirmed in the Lutheran Church, his mother gave him a telescope. This gift began a lifelong interest in astronomy and space. Soon von Braun picked up an astronomy magazine that contained an advertisement for Oberth's first book. When von Braun received the copy he ordered, he was nonplussed to discover that it was full of equations and scientific jargon rather than fantastic descriptions of space travelers. But his interest in space was so strong that he began to take his physics and math courses much more seriously as vehicles to understanding Oberth's writings. He also began experiments with rockets that resulted mainly in broken windows and trampled gardens. He finished his schooling at another boarding school on the island of Spiekeroog, where he helped build an observatory and acquired a passion for sailing.

Upon finishing school, von Braun enrolled in the Charlottenburg Institute of Technology, where he not only learned theory but the practical realities of engineering, in part through a stint as a machinist's apprentice. By this time, the efforts of Oberth and the VfR to publicize space travel were at their height. Von Braun telephoned Oberth, with whom he had previously corresponded, and volunteered his services. The young student helped Oberth with an educational display as well as with the Kegeldüse rocket engines he was testing in 1930. During this time, von Braun also learned to fly, another of his passions, and he began a lifelong friendship with Hanna Reitsch, who was on her way to becoming one of Germany's most famous test pilots of the 1930s. In 1931, he studied for a semester in Switzerland and traveled around Europe. He and a medical student friend in Zurich, Constantine Generales, built a small centrifuge out of a bicycle wheel that they used to study the effects of gravitation on mice. The two students' experiments on the effects on the unfortunate rodents of the acceleration experienced during a rocket launch ended when von Braun's landlady discovered mouse blood on the wallpaper. "These were probably the first experiments ever of this kind," von Braun later said. Generales, a Greek who later moved to the United States, said with some hyperbole that the study marked "the birth of the science of space medicine." At the end of the year, von Braun returned to the institute in Berlin, completing his bachelor's degree in 1932. He also resumed his work at the Raketenflugplatz, where he had his fateful encounter with Becker and Dornberger.

That fall, von Braun not only began his job with the German army but also his doctoral studies at the University of Berlin, which graduated him two years later. His thesis, which was kept under wraps until after the war, dealt with the many complicated problems of liquid-fuel rocket engines. With the Depression hitting Germany with full force, the army had little money for rocket research, and so von Braun worked on research grants at Kummersdorf and assembled a group of graduate students to help him at modest rates of pay.

Less than two months after von Braun started working for the army, Hitler was named chancellor of a new coalition government whose other leaders hoped to co-opt Hitler and his National Socialist Party. Baron von Braun, a conservative nationalist aligned with those who thought they could tame Hitler, retired when he was not asked to serve in the new government. Soon it became clear that Hitler could not be co-opted, and he and his Nazis established complete control over the German government.

At first, the younger von Braun had only one mechanic and a low priority for using other workers and equipment at the army facility. But the army soon hired Walter Riedel, an engineer ten years von Braun's senior, and later Arthur Rudolph, another engineer who had helped Valier with the rocket car that killed him. A newly graduated engineer named Kurt Wahmke was also employed by the army, but he died in 1934 in an ill-thought-out experiment mixing fuels. Von Braun went to work testing engines and building his first rocket, a gyroscope-stabilized vehicle called the A-1 (*A* for Aggregat or "assembly"). When the A-1 exploded in tests at the end of 1933, von Braun decided that a major redesign was in order and began work on the A-2, which was slightly longer than the A-1 at five feet, three inches. This A-2 incorporated many improvements over other rockets and was certainly the most advanced of its time. Two A-2s, nicknamed Max and Moritz after popular cartoon characters, were launched from an island in December 1934. Max flew more than sixty-five hundred feet high, and Moritz repeated the feat the next day.

By then it was becoming clear that a new testing ground was needed. Von Braun complained that he couldn't work on computations when machine-gun testing was taking place nearby, but having to move rocket testing outside the site compromised the secrecy of the work. As Hitler increased his control of the country and won popular support by strengthening the economy and enlarging the military, the army gained more resources for rocket work. Early in March 1935, Hitler formally repudiated the Treaty of Versailles and revealed the existence of the Luftwaffe in defiance of the treaty's prohibition against an air force. Early in 1935, Luftwaffe major Wolfram von Richthofen, a cousin of the Red Baron and himself a World War I air ace, visited Kummersdorf in his capacity as director of technical development of the air force. The visit heralded a period of cooperation between the two services that saw von Braun and others work on rocket-plane projects for the Luftwaffe in cooperation with the Heinkel and Junkers aircraft firms. More importantly, the two services collaborated on building the new rocket test site von Braun sought.

Nevertheless, the rivalry remained. Von Braun, who was used to an annual budget of eighty thousand marks, recalled that Becker, by 1935 a general and the head of Army Ordnance, reacted with fury when told that von Richthofen planned to spend five million marks a year on the Luftwaffe's rocket research.

"Just like that upstart Luftwaffe," Becker said. "No sooner do we come up with a promising development than they try to pinch it! But they'll find that they're the junior partners in the rocket business!"

With that Becker announced that he would get six million marks a year for the army's rocket research. This first, but by no means last, instance of interservice rivalry in the field of rocket and space research transformed the nature of German rocket research.

The Birth of Peenemünde

In December 1935, the German military began looking for a suitable testing site. That Christmas, von Braun went home and discussed his problems with his parents. "Why don't you take a look at Peenemünde?" suggested his mother, who had grown up on a nearby estate. "Your grandfather used to go duck hunting up there." Soon, officials from the two services checked out the area, a wilderness area on the island of Usedom in the Baltic Sea about 120 miles north of Berlin, and acquired the land. The Luftwaffe set up testing facilities west of the army testing site at Peenemünde East. Not yet afflicted by the bureaucratic traditions of the army, the Luftwaffe in fact constructed both sides of the facility.

To justify the spending, Army Ordnance needed specifications for the final weapon, which was already designated the A-4. Dornberger set requirements based on his experience with the Paris Gun, which could lob a shell packed with twenty-two pounds of explosive. Dornberger wanted the new weapon to be one hundred times larger, roughly a ton, and be capable of traveling twice the distance of the Paris Gun. The missile had to be transportable through railroad tunnels, and its engine had to provide fifty-six thousand pounds of thrust, many times more than the engine being built for the A-3. Historian Michael Neufeld has written in his history of the German rocket program that Dornberger's specifications for the A-4 revealed "the flawed thinking that lay behind the German missile program from the outset." Noting that the Paris Gun was a "triumph of narrow technological thinking" that had little effect on the French in 1918, the A-4 was similarly afflicted with the "lack of a well-thought-out strategic concept of how the missile could affect the course of a war." Indeed, the German army in both world wars was hobbled by its combination of tactical and operational strength and strategic incompetence, Neufeld argued. In the 1930s, Hitler's military buildup allowed the military to spend money on programs like ballistic missiles with few questions asked.

The new rocket-testing center at Peenemünde opened in 1937, although the test stands weren't ready until the following year. Soon after Peenemünde opened, the cooperation between the army and the Luftwaffe declined for a variety of reasons, including the transfer of von Richthofen to other work. The Luftwaffe developed jet aircraft and carried out its own rocket work for a time with the help of Eugen Sänger, Irene Bredt, and others. Its rocket efforts declined after the war began, however. Sänger and Bredt designed a new military version of their Silverbird winged rocket vehicle that became known as the antipodal bomber, a design that would impress Soviet dictator Joseph Stalin after the war when he sought the ability to deliver bombs across the oceans.

The army missile effort was growing and concentrating on a new rocket, the A-3, which contained a scaled-up version of the engine used in the A-2 and an advanced gyroscopic guidance system in place of the A-2's single gyroscope. In December 1937, the twenty-two-feet-high A-3 was ready for flight testing on the island of Greifswalder Oie, which lay off Peenemünde (and ironically was the place Oberth had wanted to launch his rocket in 1929 publicizing *Frau im Mond*). The launches of the four A-3s that month were almost identical failures. After a few seconds of ascent, the parachute popped out, the rocket turned into the wind and then fell back to earth and exploded after a premature engine cutoff. Investigation revealed that the control system was not strong enough to control the A-3s. The failures taught von Braun and his group important lessons about rocket systems that they absorbed as they geared up to build even bigger rockets. The jump from the A-3 to the A-4 already on the drawing boards was so great that von Braun's team decided another intermediate vehicle was needed to test systems for the A-4. Thus, work began on the A-5, which contained more advanced guidance systems and first flew successfully in the fall of 1938. Improved versions began flying a year later. Although the A-5 was a little over twenty feet long, roughly the same size as the A-3, it served as a successful test bed for the much larger A-4, particularly for various guidance and control systems. More than two dozen A-5s flew from Greifswalder Oie, some more than once after successful parachute recoveries.

On March 23, 1939, von Braun's twenty-seventh birthday, he had his first encounter with Adolf Hitler when the Führer came to Kummersdorf to see a demonstration of rocket engines. Unlike most visitors, Hitler was unimpressed by the roaring rockets and von Braun's and Dornberger's explanations of their potential, although he quizzed them closely about their use as weapons. After lunch, Hitler left, commenting enigmatically, "Well, it was grand." Hitler's visit was later deemed not to have helped the rocket program, but it didn't hurt it either. In any case, Hitler never again saw the rocket work in person.

During this time, the development team at Peenemünde continued to grow, from 411 personnel at the beginning of 1938 to triple that number in September 1939. This rate of growth continued for the next two years. Hitler's invasion of Poland in September 1939 marked the beginning of World War II and the consequent mobilization of the German economy. Military and industrial leaders jostled with one another to win priority status for their projects. Becker, Dornberger, and other leaders of the rocket program had to fight to maintain their priority for funding and for expensive materials such as steel, and to protect their personnel from the military draft. They began to plan for accelerated mass production of the A-4 in a bid to talk up its potential and win support for their program.

Hitler, who often delayed making important decisions and preferred to rule by playing people and organizations off each other, was himself forced to take a bigger hand in these resource allocation matters. The Führer had wanted to invade France in the fall of 1939 immediately after his successful conquest of Poland, but an ammunition shortage was one cause of the invasion's postponement until 1940. Becker and Dornberger's wish to obtain a top-priority steel quota to accelerate A-4 work fell by the wayside as Army Ordnance dealt with the ammunition shortage. The army also came under political attack during this period, and Hitler appointed a minister of munitions to facilitate production. At the height of this interservice rivalry in 1940, Becker committed suicide after hearing of an attack on his personal reputation. The infighting continued amongst the services and their pet projects. New priority categories were created to allow programs for aircraft, ships, or tanks to leapfrog ahead, forcing the other programs to seek these same new higher-priority designations. Though these problems primarily affected what Neufeld called Dornberger's "ill-conceived" and expensive A-4 production plant at Peenemünde, they also helped slow down development of the A-4 itself.

In spite of these political problems, von Braun's team made major progress on the A-4 from 1939 to 1942. Under the direction of Walter Thiel, the development of the A-4's engine progressed such that the problems that limited the potential of the A-3 and A-5 engines were evaded. The new engine, which was seventeen times larger than the A-3 engine, included a number of innovations in chamber and nozzle design, fuel pumps, engine cooling, and fuel mixing that improved performance and reliability. Because the A-4 would fly faster than the speed of sound, a giant wind tunnel was built at Peenemünde and a staff headed by aerodynamicist Rudolf Hermann worked on the shape of the new missile to ensure it met the challenges of flying at these new speeds. Dornberger later remembered the expensive wind-tunnel building was one of the "showpieces" of Peenemünde, complete with an inscription in the wall of the reception room: "Technologists, physicists and engineers are among the pioneers of the world."

A third area where technical breakthroughs were needed and made was in guidance. Earlier guidance work had been farmed out to industry, but Dornberger pushed to have as much work as possible on the A-4 done in house. So a new guidance group created under Ernst Steinhoff built on earlier advances to create a gyroscopic guidance system for the A-4, with heavy involvement by von Braun himself. As well, guidance beams used to direct bombers to their targets were adapted to keep the missile on course. Many of these advances were assisted by engineers in German universities, whose expertise was sought out at Becker's behest. As Neufeld has written, "only the existence of a massive and well-funded organization allowed the rocket group to create working technology in a short period of time."

As recounted by Neufeld, the relations between the Army Ordnance rocket program at Peenemünde and the Luftwaffe was a mixture of competition and collaboration. The Luftwaffe gave important assistance to the rocket work, and von Braun himself did his compulsory military service in the late 1930s in the Luftwaffe's reserve, but the airmen looked on jealously as resources were poured into the A-4. In 1942, the Luftwaffe began work on a pulse jet–powered flying bomb, the Fi-103.

The political lobbying for the rocket program continued, and on August 20, 1941, when Hitler's invasion of the Soviet Union was still going well, Dornberger and von Braun went to the Führer's headquarters in East Prussia and showed him films of successful A-5 launches and other rocket activities. This time, Hitler was impressed, saying the A-4 would be of "revolutionary importance for the conduct of warfare in the whole world" and calling for mass production of the rocket in numbers beyond what would have been possible. In spite of this, Dornberger and his allies had to continue their fight to maintain a high priority and obtain needed materials.

The A-4 Flies

Above all, they needed to get the A-4 into the air. At the time of his visit to the Führer, Dornberger was close to giving up hope that the A-4 would make its first flight before the end of the year. Manufacturing and design problems, along with mishaps that destroyed test vehicles and test stands, kept the A-4 on the ground. Von Braun and others were already doing preliminary work on more advanced rockets including winged, multistaged vehicles that could strike the United States, and incidentally send payloads into space. But Dornberger forced his rocket team to concentrate on the A-4, complaining, he later wrote, that von Braun "reveled in any project that promised to be on a gigantic scale, and usually in the distant future. I had to brake him back to hard facts and the everyday." The more advanced projects were wound down, and the only other project the team kept up was an antiaircraft missile, the Wasserfall, one of the first attempts to use then-exotic rocket fuels that ignited on contact. The mishaps that angered Dornberger continued, but on June 13, 1942, a high-level delegation headed by the new armaments minister, Albert Speer, looked on as the first A-4 launch took place. Although the flight fell well short of hopes, it impressed the delegation by flying past the speed of sound, Mach 1, before breaking up. A second launch in August also ended with the premature breakup of the rocket.

Finally, on the afternoon of October 3, the third launch was a complete success: the A-4 flew 50 miles high and splashed down in the sea 120 miles down range. One witness of this historic launch was Oberth, who had finally been called to work in the rocket program in 1941 after having pursued academic rocket studies in Vienna and Dresden since 1938. In the celebrations that followed the launch, Dornberger couldn't resist breaking his stick-to-military-business orders when he boasted that on that day "the spaceship has been born." But he added, even more prophetically: "I warn you: your headaches are by no means over, they are just beginning!"

Most of the A-4 test launches that followed weren't as successful, but the film of the October launch impressed everyone who saw it, including Hitler, who got a special showing the following July narrated by von Braun and Dornberger. The growing size and prominence of the program meant others sought to take over the A-4's production, notably Speer's munitions ministry and then Heinrich Himmler's SS (Schutzstaffel), the notorious Nazi paramilitary organization that controlled the Gestapo secret police and concentration camps. Himmler recruited a reluctant von Braun to join the SS in 1940, and finally, under orders from Dornberger, he accepted.

Although von Braun had a conservative, nationalist upbringing, he was not greatly interested in politics. He was committed to building rockets to advance his space dreams, and he was prepared to do what was necessary to help make those dreams a reality. In 1937, he joined the Nazi Party when asked to do so by his superiors, just as he later joined the SS. Many other Peenemünde engineers and scientists had attitudes similar to von Braun's, but some like Arthur Rudolph were known to be ardent Nazis. Like many military officers, Dornberger was also supportive of the Nazi regime. As for von Braun, Himmler tried to butter him up by promoting him in the SS. Von Braun is known to have worn his black SS uniform only once— during one of Himmler's visits to Peenemünde in 1943, under orders from Dornberger. When Himmler tried to enlist von Braun in his efforts to take over the rocket program, the technical director resisted.

The army had long tried to keep its rocket research as secret as possible, starting by suppressing all public discussion of its work in the 1930s. Although word began leaking about the rocket effort as war broke out in 1939, Germany's enemies ignored these early warnings. But as the war went on and Germany's successes ended in the winter of 1941 with the Soviet resistance to Hitler's invasion of Russia, followed by Allied bombing of German cities, Nazi propaganda began to warn of wonder weapons. Foreign laborers were employed in and around Peenemünde in spite of Hitler's wish to avoid their use, and these workers became a source of information for Allied intelligence services. And as more intelligence arrived in London in 1943 about the German rocket program, British leaders sent reconnaissance aircraft over Germany to photograph the buildings, test stands, and residential compounds of Peenemünde.

As a full moon shone above, a massive Royal Air Force bombing attack began shortly after midnight on August 18, 1943. The first target was the residential compound, where the RAF hoped to kill Peenemünde's engineers and scientists in their beds, but the first bombs landed too far south because of a targeting error. With the exception of Walter Thiel, Peenemünde's leading lights made it to safety, though von Braun risked his life to save precious work documents from a burning building. More than seven hundred people were killed in the raid, the vast majority of them foreign workers, and many buildings were destroyed or damaged, though not the main factory building, the wind tunnels, or the guidance building. To prevent further bombings, most damaged buildings were left untouched, and many research facilities and personnel were dispersed around Germany, which actually compounded the bombing attack's damage to the rocket program.

Among those moved away from Peenemünde was Oberth, who was already unhappy about his low-level assignments. He was sent to central Germany to work on an antiaircraft rocket.

Dora

Himmler wasted little time using the bombing as an excuse to win a bigger role in the rocket program. He argued to the Führer that further rocket production should be moved underground and the manufacturing work done by concentration camp workers, who could be cut off from the outside world. Hitler agreed, and production facilities were moved to a network of underground tunnels in the Harz Mountains in Thuringia near the city of Nordhausen. Although the tunnels had already been used to store fuel and chemicals, the underground complex had to be vastly enlarged to build A-4s and other weapons. Himmler entrusted the work to SS general Hans Kammler, an architect, engineer, and fanatical Nazi whose ruthlessness and brutality stood out even in the SS, for which he had already overseen the construction of extermination camps. Soon concentration camp prisoners from Buchenwald were shipped in cattle cars to the tunnel complex, which was known as the Mittelwerk, as well as to a new camp that would become known as Dora.

The most serious moral charges leveled against the engineers and scientists of Peenemünde relate to their participation in the use of this slave labor. Most histories of the German rocket program date the beginning of the prisoners' use in A-4 production to the inception of the Mittelwerk following the RAF bombing in August 1943. But as Neufeld has pointed out, labor shortages predating this raid had caused Peenemünde production chief Arthur Rudolph to begin using SS concentration camp inmates to work on the A-4 in April 1943. Before that, Polish and Russian prisoners of war had also been used as laborers in the area.

As the fall wore on, thousands of Polish, Russian, and French concentration camp prisoners were enlarging the tunnels of Mittelwerk and beginning missile production under unspeakable conditions. Books about Dora written by survivors Jean Michel and Yves Béon have spelled out in great detail the horrors of the camp, whose reputation was so fearsome that most prisoners believed that transfer to Dora constituted a death sentence. They described a hell on earth where inmates died of disease, starvation, overwork, and beatings from other prisoners and the truncheons of SS guards and inmate Kapos. The prisoners were housed in the tunnels amid noise, cold, and filth, and deprived of adequate food, drink, sanitation, and rest. Prisoners from different nationalities were divided against one another, and each inmate's closest companion other than hunger and pain was fear of fellow prisoners. One routine in Dora was hangings of prisoners during inmate assemblies. Michel wrote that the SS "defied the most elementary principles of humanity. They tortured, beat, strangled, dispensed slow death or brutally assassinated defenceless people." Albert Speer later professed to be horrified when he visited in December: "It was the worst place I had ever seen," the munitions minister recalled years later. "Even now when I think of it, I feel ill." The death rate at Dora rose, and soon a crematorium was erected on the site. Other weak and sick inmates were transported to death camps. Early in 1944, persuaded that rocket production was suffering because of the poor conditions, an outdoor camp was built, and conditions for the inmates of Dora improved, but only temporarily.

Meanwhile, the head of the SS, Himmler, was still working to enlarge his control of the rocket program to include the research and development work led by von Braun. In February, Himmler summoned von Braun and asked him to subsume his work under SS control and away from the bureaucracy of the army. In his account of the meeting, von Braun refused, praising Dornberger. "I ventured to compare the A-4 with a little flower that needs sunshine, fertile soil and some gardener's tending—and said that by pouring a big jet of liquid manure on that little flower, in order to have it grow faster, he might kill it." The following month, the Reichführer SS used another tool available to him when he had the Gestapo arrest von Braun, his younger brother Magnus, and two leading engineers from Peenemünde, Klaus Riedel and Helmut Gröttrup. Von Braun was accused of being overheard promoting space travel over the military needs of the Third Reich, and Riedel and Gröttrup were accused of leftist sympathies. Dornberger worked frantically to win their release. After a fortnight, when Speer interceded with Hitler, Dornberger was able to win their liberation from a detention that for von Braun would later prove to be most helpful for his reputation.

Although Himmler had again failed to gain total control of the rocket program, the political fallout from the failed assassination attempt on Hitler on July 20, 1944, caused the Führer to invest the SS with new powers that included more control over the reorganized rocket program. By then, the Red Army was moving toward Germany, and the Allies had established a beachhead in Normandy and were moving up the boot of Italy. On June 13, the first Fi-103 "buzz bomb," which the German propaganda ministry named the V-1, the *V* standing for Vergeltungswaffe or revenge weapon, was launched across the English Channel. The V-1s soon began to strike fear among the English, who knew that the sudden silencing of the cruise missile's distinctive noise at engine cutoff meant it would soon hit its target. But its slow speed and telltale buzz allowed the British to develop a set of countermeasures that ranged from barrage balloons to sending up fighters to "tip" the V-1's wings. All these measures limited its effectiveness.

In spite of numerous developmental and production problems that caused many A-4s to fail in test flights, and in particular to disintegrate late in flight, the missile, dubbed the V-2, was first fired in anger against liberated Paris on September 7. Because there was virtually no warning that a V-2 was about to strike, no defense had been prepared against it. But the weapon's impact was limited because its warhead carried less than a ton of explosives, and it would sometimes break up before impact or fail to explode until its impact had dug a hole. Standing forty-six feet high and with an engine that packed fifty-six thousand pounds of thrust, the V-2 was far and away the most advanced rocket of the war. One of the arguments that Dornberger and others had used to get priority for the V-2 was the specter of other countries such as the United States and Soviet Union having missiles of their own. As is well known, Goddard hobbled his rocket effort by refusing to work with others, and most of the Soviet rocket pioneers of the 1930s had been killed or arrested by the secret police before the war. Moreover, the Russians and other combatants used rockets as an artillery weapon. The U.S. military began using the bazooka, a small rocket grenade that could be launched by individual soldiers, along with other small solid rockets for a variety of uses. The Soviets fired Katyusha solid rockets, which were launched together in large numbers and had a range of three miles. None came close to the V-2.

As 1944 turned into 1945 and Germany's situation worsened, so did conditions at the rocket development and construction facilities. Though most of the destroyed buildings at Peenemünde weren't rebuilt to discourage further bombing, and many of the surviving facilities and test stands were moved to Poland or other parts of Germany, Peenemünde was bombed again in 1944. Last-ditch efforts on a longer-range, winged version of the V-2 and on the Wasserfall antiaircraft missile were by now under way. And the suffering at Dora and the Mittelwerk worsened. As food became scarce in Germany, the already starving prisoners were the first to feel the shortages. Although the first prisoners at Dora didn't include Jews, thousands arrived late in the war as the SS emptied concentration camps in the east before the advancing Red Army. The factory continued to build V-2s until late March, and toward the end discipline was enforced by a wave of gruesome executions inside the Mittelwerk factory in front of other prisoners. After many prisoners were killed in an RAF bombing attack on the area, thousands were shipped or marched to Bergen-Belsen, many of them murdered on the way. Of the sixty thousand prisoners who worked in the Dora camps, about a third lost their lives, and about half of them were involved in V-2 production.

In January, the sounds of Russian artillery could be heard from Peenemünde, and von Braun and the others began to ponder their futures beyond the Third Reich. One Peenemünde engineer famously said: "We despise the French; we were mortally afraid of the Soviets; we do not believe the British can afford us, so that leaves the Americans." From comments like this and the memoirs of the Peenemünde engineers, the legend has grown that they chose to deliver themselves to the Americans at war's end. But for most of the path that led from Peenemünde to the places in Thuringia and Bavaria where the rocketeers surrendered to American forces, the engineers and scientists were following orders. Kammler's order to evacuate Peenemünde came on January 31, and the order was confirmed by the head of Army Ordnance the next day. The SS remained a strong presence, and as Neufeld has pointed out, the rocket team's chances of defying Kammler's orders at the time were very small. In February, under von Braun's direction and the watchful eyes of the SS, the rocket team and any equipment that could be moved were transported to Thuringia, not far from the Mittelwerk plant. Von Braun left Peenemünde for the last time late in the month, and only a few people stayed behind. The Soviet forces didn't take Peenemünde until May 5 because they first turned south to take Berlin.

In late March, Germany's western front began to collapse, and the last V-1s and V-2s were fired. As American forces approached Thuringia, Kammler ordered von Braun and five hundred of his top people to move to Oberamergau in the Bavarian Alps. Kammler may have hoped to use the engineers to bargain with the conquering forces, and again the rocket team had no choice but to follow his orders. In early April, Kammler took his leave of the rocket team and embarked on a last-ditch crusade to stave off Germany's defeat before he disappeared without a trace. As the moment of surrender neared, American forces took the area in Thuringia where most of the rocket team remained, as von Braun's five hundred were finding lodgings in Bavaria to wait out the rest of the war. On May 2, Magnus von Braun rode his bicycle from the hotel where his brother, Dornberger, and others were staying, and arranged their surrender to American forces in the area. Hitler was dead, Germany was in ruins, and the Peenemünde rocket team, like most of their fellow Germans, faced an uncertain future.

By war's end, more than thirty-two hundred V-2s had been fired at targets in England and Europe. Although most people believed that London bore the brunt of the V-2 attacks, in reality Antwerp, the Belgian port used by the Allies to transfer weapons, supplies, and fuel onto the continent, was the major target for the weapon Dornberger in later years liked to call a "flying laboratory." About five thousand people were killed by V-2 strikes, but Neufeld estimates that more than ten thousand slave laborers died making the V-2s. The V-2, Neufeld notes, is thus unique in that more people died making it than being hit by it. The economic cost of making the V-2s was the equivalent of manufacturing more than twenty thousand bombers, yet the casualties produced by the V-2s amounted to less than many single bombing attacks. As well, the V-1, which was produced in a fraction of the time and the budget consumed by the V-2, probably created as many or more problems for the Allies. Indeed, von Braun and many of his colleagues pointed to the resources they diverted from other wartime purposes in defending their roles in the Nazi regime.

Neufeld has argued that the V-2 was a weapon ahead of its time because it was built before the availability of the precision guidance that would have met the targeting specifications that Dornberger had set for it. And it was built before atomic weapons were available as potential warheads. Hitler's anti-Semitism had forced the cream of the German nuclear physics community into exile, and Germany probably did not have the economic capacity to match the U.S. Manhattan Project that developed the first nuclear weapons in 1945, let alone develop the V-2 at the same time.

Most of the experts working at Peenemünde escaped judicial consequences for what they did during the war. Dornberger was held by the British in preparation for a trial for his role in the deaths caused in England by the V-2. But they released him when they considered the implications of making such a charge when the RAF had participated in far bloodier bombings of civilians. Two dozen low-ranking SS officers and Kapos from Dora were tried in 1947, but the proceedings yielded little information about the roles of Arthur Rudolph or von Braun in the horrors of the Mittelwerk, though both testified. A German engineer who had gone to the United States to work for the U.S. Air Force after the war was among the four people acquitted in the trial. After that, the horrors of Dora were largely forgotten until the 1960s, when some Dora survivors confronted von Braun in print in the French media. Von Braun answered their charges, but minimized his role in the activities at Mittelwerk. In 1975, a French resistance fighter and concentration camp inmate named Jean Michel wrote his memoirs of the camp, titled *Dora*. The book was published in English in the United States four years later, after von Braun's death, and just as the U.S. Justice Department was establishing an office to pursue Nazi war criminals. A law student who later worked at the office picked up Michel's book and began investigating Dora and particularly Rudolph's work. By then Rudolph was living in retirement in California, but he agreed to surrender his U.S. citizenship after the Justice Department tracked him down, and he departed for Germany in 1984. Before he died in 1995, Rudolph unsuccessfully fought to return to the United States. Despite allegations against some of the other German rocket experts, none were prosecuted.

As for von Braun, though there is strong evidence that he was not an ardent Nazi, including the fact of his arrest by the SS, documentary evidence shows that he was involved in discussions about the use of concentration camp labor, and on a number of occasions he went to the Mittelwerk facility for meetings or to deal with production problems. As former inmate Jean Michel put it: "I claim only that Dornberger, von Braun, Gröttrup and all those lumped together conveniently as the 'Peenemünde scientists' knew perfectly well what crimes were perpetrated at Dora." Von Braun's arrest by the SS is a good reminder that he and the others were not working free of compulsion at the hands of those more dedicated to Nazism. And many of those in the rocket program were motivated by the fact that these weapons would be used against countries that were laying waste to German cities with devastating bombing campaigns. Yet their participation in one of the most horrifying episodes in human history will forever taint their place in history.

The German rocket program began at a time when only Robert Goddard and a few other enthusiasts had built and launched small liquid-fueled rockets, and it ended in 1945 with a rocket whose engine delivered fifty-six thousand pounds of thrust and that had flown more than one hundred miles high. The technical innovations that made the V-2 work as a rocket if not as a terror weapon laid the foundation for a new generation of weapons as well as for space exploration programs around the world.

"Neither the V-2 nor the V-1, nor any other great technological invention of recent decades, can be associated with the name of any one man," Dornberger wrote. "The days of the lonely creative genius are over. Such achievements can only be the fruit of an anonymous team of research specialists working selflessly, soberly, and in harmony." The lonely geniuses Tsiolkovsky, Goddard, and Oberth had given way to inspired team leaders like von Braun in forging the way into space. The first rocket team had been formed to serve one of the darkest causes in human history. Many of its members would get a second chance to achieve their dreams in a country that had defeated Nazism and sought their help in another struggle that would last decades.

6. Rockets, Balloons, and the Right Stuff

The voice of him that crieth in the wilderness, Prepare ye the way of the Lord, make straight in the desert a highway for our God.

Isaiah 40:3

Five months after he and his colleagues surrendered to the United States Army in the German Alps in May 1945, Wernher von Braun, the former technical director of the German rocket team at Peenemünde, arrived by train in El Paso, Texas, in the custody of U.S. Army major James Hamill. After months of interrogation in Germany, France, England, and the eastern United States, von Braun encountered complications on his train ride from Washington DC to El Paso. Von Braun's presence in America was a secret, so the German rocketeer was not supposed to talk to other passengers or indicate his identity in any way. When he and the young major changed trains in St. Louis, however, Hamill found that they had been assigned seats in a car full of wounded veterans, who likely wouldn't take well to traveling with a leading scientist from an enemy nation. But once Hamill got them moved to another car, he found that they couldn't sit together. Though this was against orders, von Braun had already won Hamill's trust. As the train pulled into Texarkana, however, Hamill saw to his alarm that von Braun was engaged in conversation with another man. It turned out that the man had asked von Braun who he was and where he was from. The German said he was involved in the steel business in Switzerland. To von Braun's dismay, the man was familiar with both Switzerland and the steel business. Fortunately, the man had to get off in Texarkana, and he bade farewell by thanking von Braun and his fellow Swiss for helping America beat "those Germans."

For von Braun and the more than one hundred Germans who joined him in late 1945 and early 1946 at Fort Bliss in El Paso the corner of the United States they found themselves in seemed right out of the Wild West, a place very new to their experience but not wholly unfamiliar. Most Germans had grown up reading of the American west through the prose of the German novelist Karl May, who set some stories in the El Paso area. But the reality of Fort Bliss in the immediate aftermath of World War II was vastly different from May's vision of the Wild West in the nineteenth century.

The German rocketeers were hardly alone in facing new and troubling realities in the months after the atomic bombs dropped on Japan had ended World War II. Americans looked to the future with a mixture of anticipation and dread. The end of the war meant that soldiers and sailors would be coming home at last. It also meant an end to rationing and other wartime restrictions, which had dampened the economic prosperity that came with the war. But many Americans also feared that peace could mean a return to the Depression of the 1930s that had blighted so many lives and never really ended until America became engaged in the war. Japan and Germany were vanquished, but Europe and much of Asia lay in ruins. And what of America's wartime ally, the Soviet Union? Before the war, the Stalin dictatorship had long been isolated from the world, and particularly from the United States, where communism had been suppressed. Only the most diehard communists had failed to see the ruthless character of Joseph Stalin as he moved from steadfast opposition to fascism to signing a pact with Hitler carving up Poland on the eve of war in 1939. By the time the United States entered the war in December 1941, Hitler had turned on Stalin, and his armies were threatening Moscow and Leningrad. Stalin was then in the fight against Hitler, and so the Soviet Union was welcomed as an American ally, which it remained through the war. As 1946 dawned, many Americans still saw Soviet Russia as an ally, despite disquieting signs that it was planning to consolidate its control over the Eastern European countries still occupied by its troops.

As the United States moved from its embrace of peacetime pursuits to the opening of the Cold War with the Soviet Union, the technological portents of that new struggle were already taking shape in the deserts of the American west. El Paso, which occupied the westernmost tip of the state of Texas, lies south of the central portion of New Mexico, where on July 16, 1945, the

United States had exploded the first atomic bomb before using it twice the next month with devastating effect on Japan. The site of that historic blast was part of the White Sands Proving Ground, where in the coming months von Braun and his colleagues would launch the V-2 in scientific and engineering tests, along with other missiles under development in the United States. To the northwest of White Sands lay Roswell, the former home of Robert Goddard. Further to the west lay the great Mojave Desert of California, where in a desolate desert airfield, some of the best pilots in the United States were gathering to fly the fastest aircraft ever built. The atomic bomb was transforming warfare, and much of that transformation involved the missiles and jet aircraft that were taking to the skies of the U.S. southwest in the late 1940s and 1950s. Together, these efforts would eventually lead to flights of astronauts into space.

Moving Out of Germany

All of Germany was in confusion as the war ended in May. Displaced persons looked for their homes and loved ones in shattered cities and just enough food and drink to stay alive. The United States, British, French, and Soviet armies occupied portions of the country, but they would soon have to shift to occupation zones agreed to by their leaders. All were on the hunt for military and scientific experts, documents, and equipment that could be of value to their militaries, including advanced jet and rocket aircraft. On orders from the Pentagon, the chief of the U.S. Army Ordnance Technical Intelligence group, Col. Holger N. Toftoy, set up "Special Mission V-2" to go to the Mittelwerk plant—then under American occupation but due to be handed over to the Soviets—and gather the components needed to assemble and fly one hundred V-2 rockets. The officers involved in the mission had to be enterprising to carry out their orders in the face of contradictory and confusing mandates. For example, the Army Transportation Corps refused to move the V-2s for fear of offending allies, so the Ordnance Corps in effect ran its own railroad to get the rockets to port in Antwerp, a place to which many V-2s had previously traveled under their own power. Most of the bridges between Nordhausen and Antwerp had been destroyed by bombing, but officers such as Major Hamill managed to get the rockets to port just before a move by the Transportation Corps to impound all freight cars. Another problem was that no V-2 parts list existed, so soldiers simply grabbed one hundred of every available part at the V-2 plant. Once the parts arrived in Antwerp, the British got word of what was going on and tried to enlist the Supreme Allied Commander, Gen. Dwight D. Eisenhower, in a failed attempt to halt the parts shipment aboard sixteen Liberty ships bound for the United States.

Colonel Toftoy's special mission was after more than the rockets. An American officer managed to bluff a German into revealing the location of a cache of Peenemünde documents, and succeeded in having them removed from their hiding place in an abandoned mineshaft and trucked to the American occupation zone just hours before the British took over the area.

Once the German rocketeers had surrendered to American troops in May, teams from the British and American armies, along with experts brought in by the U.S. military, began questioning them. The interrogation was often repetitive and sometimes bordered on the ridiculous. The rocketeers had time to confer and decide where they wanted the discussions to go, and so they limited their cooperation in hopes of winning three-year contracts to continue their work on rockets. "We were interested in continuing our work," von Braun said, "not just being squeezed like a lemon and then discarded." The questioning and bargaining were interspersed with long periods of inactivity, which the Germans filled with inquiries into the whereabouts of their families, with lectures, and with at least one musical show. A number just traveled back home, only some of them with permission.

In May, the German rocketeers moved from the Bavarian Alps, Nordhausen, and even from Peenemünde to a new location in Witzenhausen, just inside the American zone, as the Americans debated what to do with them. In late May Undersecretary of War Robert P. Patterson wrote that as much information should be obtained from the rocketeers as possible in Germany, and that those who were brought to the United States should stay for as short a period as possible. By July, the army had set up Operation Overcast to bring about one hundred of the rocket team members to the United States. Toftoy and his officers began negotiating with the Germans and got them to agree to six-month contracts. A major issue was their families, the Americans agreeing to house them in a barracks near Landshut, near Munich. Although camp conditions were difficult, they were far superior to the conditions enjoyed by most Germans, who often had difficulty keeping themselves fed in the postwar chaos that gripped much of Europe.

After questioning in England, von Braun and six others flew to the United States in late September, helping the Americans process V-2 documents at the Aberdeen Proving Ground in Maryland, where not long before Goddard had worked. Although he was suffering from hepatitis and the effects of an arm badly broken in a car accident prior to his surrender that spring, von Braun was the first to go to Fort Bliss in El Paso. He was followed in the next few months by a total of 118 of his colleagues from Peenemünde. The V-2s by this time had arrived in White Sands, and early in 1946, Overcast was renamed Operation Paper Clip. The Germans began to settle into life at Fort Bliss and called themselves "Prisoners of Peace."

They were put to work assembling V-2s, preparing them for launch, and training army crews to launch them. But in the postwar atmosphere of demobilization and a return to civilian pursuits, there was little money available for or interest in building new rockets, let alone spacecraft. Moreover, not all the required parts were available or in condition for use, and funds for replacement parts were scarce. The rocketeers lived in spartan wooden buildings converted from a hospital at Fort Bliss, and were allowed out only under escort since they were not even officially in the United States. Though their initial six-month contracts were renewed, their future seemed uncertain, and many spent their leisure hours making furniture, learning English, or in the case of von Braun, preparing a plan to fly humans to Mars. They also prepared documents on their work at Peenemünde. These reports found their way into the hands of Americans in government and industry who wanted to develop new American rockets. As a result, some of the Germans even went to visit major U.S. aviation companies to give advice on rocket and jet engines.

In late 1946, conditions for the Germans began to change. In November, the first members of their families were allowed to join them at Fort Bliss. In December, the government decided to make the presence of the Germans public. When an El Paso newspaper reporter interviewed von Braun and a few other leaders of the group, Walter Riedel's complaints about American cuisine, notably its "rubberized chicken," garnered some minor controversy. On January 16, 1947, von Braun gave his first public address in the United States, speaking about the future of rockets to the El Paso Rotary Club. His talk earned a standing ovation. In March, von Braun was briefly back in Germany, where he married his eighteen-year-old cousin, Maria von Quistorp. Late in 1948, their first daughter was born at Fort Bliss.

Although the U.S. Army's ultimate plans for the Germans remained unclear, to the frustration of von Braun and his men, the Germans' value increased as the Soviet Union moved to consolidate its positions in Europe. These moves transformed the Soviet Union from America's wartime ally to its Cold War adversary. The army used arguments about the rocketeers' security value to convince a wary J. Edgar Hoover and the Federal Bureau of Investigation to allow them to stay. But the Germans were subject to investigation and surveillance in America even as many officials ignored some of their wartime activities back in Germany. Finally, in 1948, the Germans were given a legal basis for their residency in the United States. Boarding streetcars in El Paso, they went to the U.S. consulate across the border in the city of Juárez, Mexico; obtained visas to legally enter the United States; and then returned to El Paso.

The first V-2 launch from the White Sands Proving Ground in New Mexico took place on April 16, 1946, but the flight was cut short when a fin fell off. The second launch, on May 10, was a success. From these launches to the last V-2 launch on September 19, 1952, sixty-four V-2s were sent aloft at White Sands. In addition, two were launched from a new launching ground at Cape Canaveral in Florida, and another was fired from the deck of an aircraft carrier. A V-2 launched in 1947 went off course when its gyroscope failed and landed next to a cemetery in Juárez, Mexico. Within minutes, concerned phone calls came in to Fort Bliss from General Eisenhower, then the army chief of staff, and George C. Marshall, the secretary of state. Just as quickly, stands selling "souvenir" pieces of the V-2 opened up for business in Juárez.

Under the coordination of a V-2 Upper Atmosphere Panel that included representatives from universities, industry, and the military, V-2 rockets flew experiment packages in their nose cones that made the first measurements of the composition and pressure of the upper atmosphere and the lower ranges of space. Instruments probed the composition of the ionosphere and the strength of radiation high above the earth. Some rockets carried instruments that looked at the sun or photographed the earth from above the atmosphere. Many of the experiment packages were placed in nose cones that could separate from the rocket and be recovered by parachute, although the recovery systems didn't always work. Eight of the V-2s sported WAC Corporal missiles serving as their second stages as part of "Project Bumper." As America's first two-stage rockets, these rockets gave the U.S. rocket ef-

10. The Bumper V-2 was the first missile launched from Cape Canaveral on July 24, 1950. (NASA photo)

fort experience in staging technologies, and in 1949 one Bumper set an altitude record of 244 miles.

Inevitably, some people began to think about the day when humans would step aboard rockets bound for space, and so some V-2s were allocated for flights of animals as a preliminary to human flights. Under the Air Force's Project Blossom, the Aeromedical Laboratory at Wright Air Development Center in Ohio was asked in 1949 to supply "simulated pilots" for upcoming Blossom launches. Dr. James P. Henry directed the animal program with the assistance of another physician, air force captain David Simons. The first animal launch was scheduled for June of 1949, giving Henry and Simons just two months to find a suitable test candidate and design a capsule. The two researchers quickly decided that a rhesus monkey was best suited to the task. Holloman Air Force Base in New Mexico had manufactured a capsule for holding one of the monkeys that would slide into a small section of the V-2's nose cone. This pioneering series of primate flights was code-named "Albert"—the name also given to the monkey selected for the first test.

On June 11, 1948, just forty-five minutes before launch, the eight-pound monkey was anaesthetized, and electrocardiograph electrodes were sutured into place to monitor his heart beat and respiratory rate. He was placed on a felt padded seat; chest straps were put in place and then secured. Next, Albert was carefully inserted into the capsule—later described by Simons as being so small that "his head had to be placed in a cramped, forward position with the neck acutely flexed." The capsule was sealed and flushed with oxygen, ready to be lifted to the top of the gantry for insertion into the rocket's upper section. Fifteen minutes later Albert's capsule was secured in the detachable nose cone at the top of the V-2.

Blossom 3, the first with live passengers, launched right on schedule, but heady success soon turned to abject failure. After reaching thirty-seven miles of altitude the nose cone separated from the V-2 as planned and began its descent. The single parachute deployed, but failed to inflate due to the heavy load and thin air. It eventually billowed out at twenty thousand feet, but the sudden shock load caused the chute to explode in shreds, sending the nose cone hurtling back to earth in a flat spin. On impact it skidded across the terrain and burst open. It was later revealed that Albert had had difficulty breathing prior to launch, and the instruments used to transmit respiratory data had also failed before liftoff. Indications suggested that poor Albert had probably died of suffocation before his historic flight even began.

Nevertheless, on 14 June 1949, a six-and-a-quarter-pound rhesus named Albert II was carefully strapped into his redesigned and enlarged capsule, which in turn was inserted into the nose cone of his V-2. The instruments worked well during the flight, but once again the parachute system failed, and the descending nose cone broke free of its shroud lines, slamming into the ground 340 seconds after takeoff. It hit with sufficient force to make a crater ten feet in diameter by five deep, burying some pieces nearly twelve feet into the desert soil. Still, this flight marked an advance over the previous flight since Albert II's rocket had reached an altitude of eighty-five miles, and the monkey had survived right up until impact. Following this flight Captain Simons regretfully left the project to take on career training as a flight surgeon, but his frustration with Project Blossom was evident. "Only the recovery" he said, "of a live animal showing no demonstrable ill-effects will permit the claim that no major difficulty has been overlooked."

Albert III was sent aloft on September 16 of that year, but once again the flight ended in disaster. Eleven seconds after liftoff, a small but violent explosion took place in the tail section of the V-2. The rocket continued to move skyward, but fourteen seconds later a second explosion occurred, and this time the missile was blown to pieces.

On December 12, another V-2 carrying Albert IV performed flawlessly. With greatly improved instrumentation and telemetry, scientists were able to accurately monitor the monkey's pulse rate, heartbeat, and respiration, and were quite satisfied that spaceflight was survivable. Sadly, however, the flight ended as before; parachute failure meant the capsule plummeted to earth, and the test monkey died on impact. One final V-2 test flight remained, but this time a mouse was sent aloft instead of a monkey, and a movie camera was used to record his movements while weightless. Though the scientists had been hoping for better results, once again the nose cone parachutes did not operate properly.

With the last of the allocated V-2s now used up, animal flights were shifted to the newly developed Aerobee sounding rockets, designed to explore the upper atmosphere. Far lighter and less ponderous than the V-2, their smaller size also compelled a redesign of the capsule. The first Aeromed Aerobee launch took place from White Sands on April 18, 1951, with Albert V on board, reaching an altitude of thirty-five miles. To the scientists' dismay, despite many months of modification work, the parachute system failed once again.

During the next flight on September 20, the single parachute worked adequately, bringing the nose cone back to earth with a solid bump after reaching a height of forty-five miles. Albert VI and all eleven mice on board survived the flight and landing. Two of the mice had been in a special transparent drum, where they were filmed floating about throughout the flight. Unfortunately, the nose cone was not recovered quickly enough from the fierce desert sun, and despite quick attention the monkey died from heat prostration soon after he was released from his cramped capsule.

The parachute designers at last came up with what they hoped was the solution to their problem—a double parachute. The first test of the new parachute system occurred on May 21, 1952, when two Philippine macaque monkeys named Patricia and Michael were launched in an Aerobee rocket together with two mice. This time everything worked well; the launch was successful, the Aerobee reached a height of thirty-nine miles, and the new parachute system worked without a hitch. The animals were recovered in good health and spirits; a great milestone in spaceflight had finally been achieved.

Breaking the Sound Barrier

Humans, not just animals, were also being subjected to new speeds and acceleration forces in the first decade after the war, sometimes with similarly fatal results. At the time of Project Blossom, humans were flying in revolutionary new aircraft whose forms were based on the major advances in aircraft design and engineering that took place in anticipation of and during World War II. Aircraft designers in America, Britain, Germany, and Russia were scrambling to build aircraft that incorporated improvements in engine and airframe design. In the United States, much of this work was done by the National Advisory Committee for Aeronautics (NACA), a government agency set up in 1915 that conducted research on aerodynamics and aircraft engines in laboratories in Virginia, Ohio, and California. The results of NACA's work were shared with the growing U.S. aircraft industry, which developed innovations of its own. Aircraft built with these advances played a crucial role in helping American and allied airpower defeat Germany and Japan.

Just a few days before Hitler's invasion of Poland opened hostilities, the first jet aircraft took to the air in Germany. Yet the overwhelming majority of the aircraft that flew in World War II were powered by propellers, and jet aircraft did not become widespread until after the war. Erich Warsitz flew that first jet on August 27, 1939, in an aircraft designed by Ernst Heinkel, with jet engines designed by Hans von Ohain, a graduate of Göttingen University. Warsitz had earlier distinguished himself by flying the first rocket planes in collaboration with Heinkel and Wernher von Braun. But Ohain's new jet engine was more immediately promising for military use because unlike rockets, which carry both fuel and oxidizer, the jet engine drew its oxidizer from the air. The German government was interested in jet-powered aircraft, and both the Heinkel and Messerschmitt companies set to work on building aircraft. The only jet aircraft deployed in any significant numbers during the war was the Messerschmitt-262 jet fighter, which entered combat in July 1944. Although it flew at 541 miles per hour, faster than any other plane in the sky, and scored success against Allied bombers and fighters, the Allies devised strategies to destroy the Me-262 on the ground or exploit its poor maneuverability.

In 1941, a British aircraft had flown with a jet engine designed by Frank Whittle, whose work on jets in the 1930s had been slowed by government indifference. Although both British and American aircraft firms began working with jet engines, Britain's Gloster Meteor was the only Allied jet aircraft used during the war. The reason, as T. A. Heppenheimer has explained in his history of flight, was that American and British aircraft were able to fly higher and faster than their adversaries using supercharged piston engines. The Germans did not have the supercharged engines, and so tried to leapfrog American aircraft such as the P-38 Lightning and the P-51 Mustang with the Me-262. When the war ended, both the American and British aircraft industries began to look to the future, and jet engines were central to it. Because jets involved high speeds and technologies similar to rocket engines, the first generation of American engineers and pilots who would build and fly rockets and spacecraft learned many important lessons with jet aircraft.

The high speeds that jet engines and rocket engines offered also presented new challenges for aircraft designers, not the least of which was the so-called sound barrier. During the war, when aircraft approached the speed of sound, which varies with altitude, they tended to lose control. Although bullets and the V-2 rocket regularly flew faster than sound, aircraft designers knew that designing aircraft that could fly safely at those speeds with jet engines and rocket engines would be difficult. The wind tunnels of the time weren't up to the task, so the U.S. military ordered that a bullet-shaped rocket plane known as the X-1 be developed to break the sound barrier. The dangers presented by the buffeting that afflict aircraft near the speed of sound were dramatized in 1946 when the famed British test pilot Geoffrey de Havilland, the son of the famous aircraft designer, died when his aircraft broke up just below Mach 1, the speed of sound.

The X-1's home base was established a few hundred miles west of the White Sands rocket launching sites, in the desolate Mojave Desert of California, one hundred miles northeast of Los Angeles. There, pilots from the U.S. Army Air Force were beginning to test new aircraft at a dilapidated military base known as the Muroc Army Air Field. The name came from the first settlers in the area, who were named Corum. Shortly before World

War II, the area and the adjacent Rogers Dry Lake, its name a corruption of "Rodriguez" from a local gold mining company, were put to use as a bombing range. The range, which was used to train bomber crews, came complete with a mockup Japanese battle cruiser known as the Muroc Maru. Also during the war, Muroc was used to test the first U.S. jet fighter, the classified Bell xp-59A Airacomet. The base offered near perfect flying weather year round, and the Rogers Dry Lake offered plenty of room for errant aircraft to land. These attributes, plus its isolation, inspired the Army Air Force to move its advanced aircraft programs after the war to Muroc from Wright Field in Dayton, Ohio.

Though the military brass liked the new site, those who had to live there had different views of Muroc. Glennis Yeager, the wife of test pilot Chuck Yeager, described her first visit to Muroc this way: "We went by the guarded gate and drove for miles and miles and there was absolutely nothing there but scrub and lakebeds and Joshua Trees. Twenty-eight thousand acres of nothing. The place was used during the war for practice bombing and I could see why. I said to him: 'Where's the base?' Chuck laughed. 'Hell, this is the busy part.' Well, there really wasn't much to Muroc in those days—a few hangars and buildings shimmering in the sun. The wind never stopped howling. The desolation took your breath away."

Capt. Charles E. Yeager was a young fighter pilot from the hollers of West Virginia whose record in World War II included twelve and a half kills and successful evasion of capture after being shot down over France. He had impressed his superiors at Wright Field enough to be accepted as a test pilot and then handed the coveted assignment of flying the X-1.

The X-1 was made by the Bell Aircraft Co. of Buffalo, New York. Its power plant was an xlr-11 four-chambered rocket engine fueled with ethyl alcohol and liquid oxygen that gave a total of six thousand pounds of thrust. The engine couldn't be throttled, but the pilot could choose to fire the four chambers singly or in groups. Each time a chamber or chambers were lit, the pilot got a "kick in the pants." Through 1946 and the first half of 1947, the X-1 was tested by Bell test pilots, but mainly by Chalmers "Slick" Goodlin. When Goodlin demanded payment of $150,000 to go beyond Mach 1, however, the U.S. Air Force and NACA, which also supported the X-1 program, decided to use their own pilots. The air force selected Yeager to break the sound barrier. Yeager saved the government money by flying on his small

11. The Bell Aircraft X-1 in flight in 1947. The shock-wave pattern in the exhaust plume is visible. The X-1 series aircraft were air-launched from modified Boeing B-29 Superfortress bombers. Air force pilot Chuck Yeager broke the sound barrier for the first time while flying an X-1. (NASA photo)

salary, and because he was only posted to Muroc temporarily, his wife and children were not entitled to housing or even medical services at the base. They came anyway and lived in ramshackle accommodations off base.

In August, Yeager began to learn the ropes of the X-1 with three flights that commenced when the rocket plane was dropped from its B-29 host aircraft. On August 29, Yeager finally lit the X-1's engines, and over the next few weeks he flew the bullet-shaped aircraft seven more times at speeds below Mach 1. Two nights before Yeager was due to attempt to break the sound barrier, he broke two ribs when he was thrown off a horse. But with his ribs taped up, he was determined to try. So on October 14, 1947, the B-29 dropped Yeager and the X-1 at twenty thousand feet, and Yeager lit two of the XLR-11's chambers. When the orange aircraft, named *Glamorous Glennis* in honor of Yeager's wife, rose to nearly forty thousand feet, he leveled off and lit another chamber. After quickly passing through turbulence below Mach 1, Yeager smoothly broke the sound barrier, and for the first time a sonic boom rumbled over Muroc. After ten minutes, Yeager shut down the X-1's engines and landed the aircraft safely on Rogers Dry Lake.

"After all the anticipation to achieve this moment, it was a let-down. It took a damned instrument meter to tell me what I'd done," Yeager wrote later. "There should've been a bump in the road, something to let you know you had just punched a nice clean hole through that sonic barrier. Later on, I realized that this mission had to end in a let-down, because the real barrier wasn't in the sky, but in our knowledge and experience of supersonic flight."

12. Air force captain Charles E. Yeager, standing in front of the X-1A supersonic research aircraft, became the first man to fly faster than the speed of sound in level flight on October 14, 1947. (U.S. Air Force photo)

Although the flight took place a month after the creation of the United States Air Force out of the Army Air Force, the only celebration of Yeager's feat was a private one at his home. The flight was kept secret for two months for security reasons before the news was finally leaked. On his next flight in the X-1, Yeager was placed in great danger when his electrical system failed at the moment the plane was dropped from the B-29. Despite such risks, he and other air force and NACA pilots continued to fly the X-1 higher and faster. In a series of flights that went on for another four years, the X-1 and its test pilots demonstrated characteristics of supersonic aircraft that were used to build a whole new generation of aircraft that flew faster than sound. Modified versions of the X-1 carried on this research until the last flight in 1958, which featured a NACA pilot named Neil Armstrong at the controls. The U.S. Navy also brought its research aircraft to the Mojave, including the Douglas D-558-I Skystreak and the D-558-II Skyrocket. In November 1953, NACA pilot Scott Crossfield became the first man to fly at Mach 2, twice the speed of sound, in the Skyrocket. Yeager eclipsed him, however, just before the fiftieth anniversary of the Wright Brothers' historic flight of December 1903 when he flew the X-1A at Mach 2.4—and nearly lost his life when the aircraft went into a dangerous spin.

13. The Bell Aircraft X-2 drops away from its Boeing B-50 mother ship in this 1955 photo. Lt. Col. Frank "Pete" Everest piloted the X-2 on its first unpowered flight on August 5, 1954. Everest made the first supersonic X-2 flight on April 25, 1956, achieving a speed of Mach 1.40. In July, he reached Mach 2.87, just short of the Mach 3 goal. (NASA photo)

These aircraft were followed by other rocket planes, including the X-2, an ill-fated aircraft that used exotic alloys in its structure and the XLR-25 throttleable rocket engine, developed by Robert Goddard's former team at Curtiss-Wright. Though the twin-chambered engine was the first throttleable rocket engine to be rated to fly humans, its complexities slowed the troubled X-2 program. Two X-2s were built. One was lost in an early test flight, but the second went on, after a shaky development program, to make history. In September 1956, Iven Kincheloe took the X-2 to a record altitude of 126,000 feet, nearly twenty-four miles high. Later that month, in the X-2's thirteenth and final powered flight, Milburn Apt set a speed record of 2,096 miles per hour, nearly Mach 3.2, but lost his life when the X-2 spun out of control shortly after he cut the engine. The X-3, the so-called flying stiletto, was another troubled aircraft that nevertheless supplied important data about flying characteristics at high speeds.

In the early 1950s, the air force's entire flight test division moved out to Muroc. In 1950 the base had been renamed Edwards Air Force Base after Capt. Glenn E. Edwards, a test pilot who died two years before in the crash of his Northrop YB-49 Flying Wing, a troublesome and exotic aircraft that was a forerunner to the stealth aircraft of the 1990s. With the brand-new jet and rocket aircraft coming to Edwards, the desert airbase became the place to go for pilots who wanted to prove they were better than the rest, possessors of what writer Tom Wolfe dubbed "the right stuff." As Wolfe wrote in his book of the same name, "You couldn't really keep a hot, competitive pilot away from Edwards." Only one place rivaled Edwards, and that was the Patuxent River Naval Air Station in Maryland, where the Navy wrung out its new fighters. Though naval aviators boasted that they were superior to air force pilots because they had to learn how to land jets on the short runways of heaving aircraft carriers, it remained true that the hottest planes of the jet age first flew over Edwards. As the Cold War intensified, notably with the outbreak in 1950 of the Korean War and its aerial duels between jet aircraft developed by the United States and the USSR, the demand for faster and more dangerous aircraft seemed insatiable. Every one of those new aircraft needed to undergo flight testing, and some would be found wanting. The taste of speed and altitude that these jets gave their pilots helped create appetites to fly faster and farther. On their way to the stars, the first generation of Americans who would fly into space passed first through Edwards or Pax River.

Rocket Sleds and Balloons

Though the test pilots of Edwards and Pax River became famous for their exploits in jets and rockets, another air force officer involved in research would make his own mark by almost single-handedly running a program that brought Americans to the edge of space using the first technology to bear humans aloft. His name was John Paul Stapp, a physician who put his own life and health on the line in order to make flying at high altitudes and high speeds safer. In 1946, the thirty-six-year-old Stapp was testing oxygen breathing systems in high-altitude aircraft at the Aeromedical Laboratory at Wright Field, when he learned the lab was about to begin tests on sudden deceleration. Stapp volunteered for the work, which began the following year in the desert at Muroc on a special two thousand–foot track equipped to carry a special rocket sled that could be brought to a rapid stop. After a series of tests with dummies to prove the rocket sled and its forty-five-foot-long braking system, one of the most powerful ever built, Stapp himself volunteered for human runs. Before his twenty-ninth and last rocket sled run eight years later on another track at Holloman Air Force Base in New

14. Photo of U.S. Air Force colonel (Dr.) John Paul Stapp. (U.S. Air Force photo)

Mexico, Stapp would make his name on the sleds. He earned the title of the "fastest man on earth" when on his last ride he accelerated to 632 miles per hour in just five seconds before coming to a bone-jarring stop in just 1.5 seconds—briefly subjecting himself to more than 40 g's. During his rocket-sled rides, Stapp suffered broken ribs, hemorrhages in his eyes, a concussion, an abdominal hernia, a fractured coccyx, and a shattered wrist. When asked to describe his rocket-sled rides, he said: "It's like being assaulted in the rear by a fast freight train." While doing this research, Stapp helped make a major contribution to popular wisdom. One of his assistants, Capt. Edward A. Murphy, cursed out a technician for making an error in preparing a rocket sled. After Stapp repeated the story to the media, Murphy's complaint became famous as Murphy's law: "If something can go wrong, it will."

Every pilot and astronaut who has had to eject or face the possibility of high-speed ejection is in Stapp's debt for his pioneering work. But even as the air force doctor ended his rocket work, he still had more ideas for testing pilots and aircraft. During a sojourn back at Wright Field between his work at Edwards and Holloman, Stapp realized that future pilots would be contending with more than just the deceleration forces he faced in the rocket sleds. Stapp knew that the problems of flying at high altitudes and in space would soon require the attention of aviation medicine. The most effective way to test human responses to high altitudes was, he concluded, with balloons, since they could remain at high altitude for long periods, unlike the aircraft of the time.

Before World War II, in a preview of the space race later in the century, balloonists from Europe, the Soviet Union, and the United States had competed to set altitude records. Balloonists in the nineteenth century had risked and in some cases found death as they ascended into the thin air of the upper atmosphere above 30,000 feet. In three flights in 1927, Capt. Hawthorne Gray of the Army Air Corps twice reached an altitude of 42,000 feet, but he died on his third flight from the effects of cold and thin air. In 1931, Auguste Piccard, a Swiss physics professor and veteran balloonist, built a fully sealed gondola for his balloon and ascended into the stratosphere from Germany with an associate, Paul Kipfer. They reached a record altitude of nearly 52,000 feet and most importantly, made it back to earth safely. Piccard repeated the feat the following year, flying slightly higher. Soviet balloonists soon eclipsed his record flights. In September 1933, a Soviet balloon, the *Stratostat*, safely flew more than 62,000 feet high, carrying Red Army commander Georgi Prokofiev, Ernst Birnbaum, and Konstantin Gordunov. The following January, another three men, aboard the *Osoaviakhim I* balloon, lost their lives when the balloon descended too quickly after reaching an altitude of 72,000 feet. The men were given hero's funerals and interred in the Kremlin Wall.

Meanwhile, the Americans were also entering the competition. On November 20, 1933, on their second try, U.S. Navy lieutenant commander Thomas "Tex" Settle and Marine Corps Reserve major Chester Fordney flew to 61,000 feet in the *Century of Progress* balloon, named after that year's world's fair in Chicago. The following July 28, the *Explorer I* balloon sponsored by the Army Air Corps and the National Geographic Society flew Capt. Albert Stevens, Capt. Orvil Anderson, and Maj. William Kepner to nearly the height reached by Settle and Fordney. As Americans listened to dramatic radio broadcasts from the balloon's gondola, the crew leaped from the balloon during its troubled descent and parachuted to safety. In October, Jean Piccard, Auguste's twin brother, and an émigré to the United States, flew the *Century of Progress* to an altitude of 57,000 feet with his wife, Jeanette Piccard, who became the first woman to travel to the edge of space.

The balloon flights were capped in November 1935 when Stevens and Anderson took *Explorer II* to a record altitude of 74,000 feet, where they reported being able to see the curvature of earth. After their safe return to earth, they held the altitude record for more than two decades.

At Holloman in the mid-1950s, Stapp set to work on balloon flights with human passengers in a project known as Manhigh. To fly the balloons, Stapp, by then a colonel, recruited two fellow air force officers he had worked with in his rocket-sled effort: Capt. Joseph Kittinger, who had flown his T-33 alongside Stapp's rocket sled on its runs down the Holloman track, and Maj. David Simons, who had served as Stapp's flight surgeon on his rocket-sled runs and had been involved in the first animal rocket flights in Project Blossom. To build the balloons for Manhigh, Stapp called on a firm called Winzen Research, headed by Otto Winzen, who after emigrating from Germany to the United States in 1937 had built a successful plastic and paper milk-container business. Winzen's passion was plastic balloons, which he was building for the air force and navy. An essential part of his success was his wife, Vera Winzen, who provided needed management skills for the firm and became known as a skilled balloon builder in her own right. By the 1950s, their balloons had been used to carry scientific equipment to high altitudes, and both the navy and air force hired Winzen Research to build balloons to carry humans.

In November 1956, the Navy balloon *Strato-Lab I* with Malcolm Ross and Lee Lewis on board, broke *Explorer II*'s record with a flight to 76,000 feet. Meanwhile, Stapp, Winzen, and the Manhigh team were developing the balloon and tiny sealed capsule for their flights. Stapp selected Kittinger to ride the balloon on a test flight, and on June 2, 1957, *Manhigh I* ascended from an airfield near St. Paul, Minnesota. Kittinger soon found that his radio wouldn't work and that an incorrectly installed valve was venting most of his precious oxygen supply outside the capsule. Nevertheless, *Manhigh I* ascended to 96,000 feet, an altitude that had only been previously exceeded by the X-2 rocket plane. Despite his problems, Kittinger kept in touch with earth using Morse code. When Stapp ordered Kittinger to begin descending, Kittinger typed out "C-O-M-E A-N-D G-E-T M-E," causing concern that he was suffering from breakaway phenomenon, a feeling of isolation and detachment. But Kittinger soon signaled his intention to descend and made it back to the ground safely.

Simons took his turn in the gondola on August 19, rising to an altitude of 101,000 feet. He was awestruck by his view of the sunset from his perch nearly twenty miles high, and when night fell he said, "I have a ringside view of the heavens—it is indescribable." The next morning, he recorded: "I felt as if I no longer belonged to the earth on this morning. My identity was with the darkness above." Simons made scientific measurements as *Manhigh II* floated for nearly a day and a half, and after dodging storm clouds and dealing with minor equipment failures in his cramped capsule made it back to earth safely. The Manhigh and Strato-Lab pilots had given humans a taste of life at high altitudes, and although their exploits had gained great publicity, they had little support from the government. Indeed, Winzen Research had to invest its own money to keep Manhigh going.

But another race to high altitudes would soon energize and overshadow the U.S. balloon programs, and much else. Although the Soviet Union had shaken American complacency when it successfully tested an atomic bomb in 1949 and a hydrogen bomb in 1953, it was still taken for granted that the United States enjoyed a healthy lead in missiles and in many other fields of technology. As Simons flew so high above the earth that August, a team of Soviet engineers was putting the finishing touches on a missile that would put an end to that complacency.

7. Korolev and the First ICBM

Our mission is to ensure that Soviet rockets fly higher and faster than has been accomplished anywhere else up until now. Our mission is to ensure that a Soviet man be the first to fly in a rocket. And our mission is to ensure that it is Soviet rockets and Soviet spaceships that are the first to master the limitless space of the cosmos.

Sergei Korolev, 1955

Late in the evening of June 27, 1938, Sergei Korolev heard the knock on his Moscow apartment door heralding his arrest by the Soviet secret police. His incarceration marked the beginning of an odyssey that nearly cost him his life and left him with lasting physical and psychological scars. Korolev was just one of the estimated 4.5 to 5.5 million people who were caught up in the Great Terror that engulfed the Soviet Union in the 1930s. The RNII rocket research institute, which in 1937 was renamed the Scientific Research Institute No. 3 or NII-3, was squarely in the crosshairs of the secret police, then known as the NKVD. The institute's sponsor, Marshal Mikhail Tukhachevsky, and its director, Ivan Kleimenov, were executed by the NKVD, along with many others. Korolev and others such as Valentin Glushko escaped the executioner's bullet but were sentenced to long stretches in the gulag, the Soviet system of "corrective camps" that almost equal Hitler's concentration camps in the annals of oppression. Though Korolev would later be favorably compared with the German rocket leader Wernher von Braun, his life in the late 1930s and early 1940s resembled the lives of the unfortunate inmates of Dora more than of the German scientists who designed the rockets they built. It was only after the war that Korolev rose to become von Braun's peer as a leader of a space program.

The Soviet rocket researchers had come under suspicion because of their links to Tukhachevsky, who had been accused of being part of an "anti-Soviet Trotskyite conspiracy." Denunciations, whether given willingly or under compulsion, played a key role in propelling the machinery of state terror. Korolev was arrested after being denounced by Kleimenov; his deputy, Georgy Lagemak; and Glushko while in the custody of the NKVD. No doubt this denunciation contributed to the strained relationship between Korolev and Glushko that afflicted the Soviet space program in the 1960s. After losing out in technical disputes with Kleimenov and then losing his job, Andrei G. Kostikov became the NKVD's chief informant at the institute during the purges. But though his actions enabled him to gain the top job at the institute, he was arrested and purged in turn during World War II.

Korolev, for his part, was accused of disruptive activities that slowed the work of the research institute. Three months after his arrest, and following torture sessions in Moscow's notorious Lubyanka prison that extracted a confession, Korolev was tried and sentenced to ten years in a labor camp. Believing that his arrest was a bureaucratic mistake, Korolev wrote many appeals to higher authorities, including Joseph Stalin himself, against the charges. Like many Soviet citizens of the time, Korolev could not bring himself to associate the Soviet leader with the terror. For many years, in fact, the terror was known as the "Yezhovshchina," after the head of the NKVD, Nikolai Yezhov, who himself became a victim of the purges in 1938. Yezhov's fall played a role in Korolev's fortune, because the new chief of the NKVD, Lavrenti Beria, wanted to burnish his image upon taking office. He chose Korolev's case to show off his own human attributes, which would rarely be exhibited again during his fourteen years as Stalin's chief executioner. In June 1939, the NKVD agreed to retry Korolev on reduced charges, but unfortunately Korolev was already on his way from prison to a gulag camp in the far eastern reaches of Siberia. Korolev worked for several months in a gold mine in the Kolyma area in one of the worst of the gulag camps, until word reached him of his retrial. The poor diet, hard work, and strict discipline, combined with the Kolyma camp's harsh arctic conditions, claimed the lives of most of the people imprisoned there. Late that year, Korolev was finally sent back to Moscow, but he had already sustained injuries and lost most of his teeth. When he got back to Moscow, Korolev received a reduced sentence of eight years, which he would not have to serve in the labor camps.

Although Korolev had hoped to be released and continued to appeal his sentence, he was sent to a new prison, this one far more comfortable than Kolyma. So many members of the Soviet scientific elite had been caught up in the terror that the NKVD created a special network of prisons for them, known by the prisoners as the *sharashka*, a Russian expression for a sinister enterprise. Among those already working in the *sharashka* were Glushko and famed Soviet aircraft designer Andrei N. Tupolev, who specifically requested that Korolev join his group working on aircraft projects in prison. Thus in September 1940, Korolev joined Tupolev's *sharashka* in Moscow, also known as Central Design Bureau No. 29 or KB-29. Although still in state custody, the engineers and scientists enjoyed a better diet than most prisoners and were allowed visits from their families.

Stalin had forestalled war in 1939 by striking a deal with Hitler that was breathtaking in its cynicism even for Stalin: between them, Germany and Russia would carve up Poland and other parts of eastern Europe. After Hitler conquered most of the rest of Europe in 1939 and 1940, however, he began looking east, and on June 22, 1941, he ended his alliance with Stalin by invading the Soviet Union. In the early months of his invasion, Hitler reached the gates of Moscow and Leningrad. The *sharashka* prisoners were moved out of harm's way, in Korolev's case to Omsk, and then in 1942 to Kazan, where Korolev found himself working in a group headed by his former deputy and accuser, Glushko. There the imprisoned engineers worked on rocket engines for aircraft. In the summer of 1944, Glushko, Korolev, and their associates were officially released by the Soviet government, but in reality little changed. Although management of the group was shifted from the secret police to the aviation ministry, the engineers continued to work in the same facility in Kazan, far away from their families. The war was still on, and in a country where more than twenty million lost their lives, everyone was expected to make sacrifices until the final victory over Germany was won in May 1945.

Churchill's Rocket Request

Stalin's terror of the 1930s had weakened Soviet rocketry efforts, which were folded into aircraft work once the war began. One notable exception was the development and flight of a rocket plane, the Bereznyak Isayev (BI), which flew seven times in 1942 and 1943. By war's end, the Soviets were seeking information on German rocket efforts. Ironically, the first notice they were given of the German effort came from none other than Winston Churchill, the prime minister of Great Britain, who, writing Stalin in summer 1944 about the German rocket effort, asked his permission to allow British experts to visit a German test range in Poland that had recently been taken by Soviet forces. Stalin responded by sending a team of Soviet rocket experts, including Korolev's old GIRD comrades, Mikhail Tikhonravov and Yuri Pobedonostsev. The Germans had removed most evidence of their rocket work before they surrendered the territories, but the Russians were able to gather pieces of the V-1 and V-2 and associated equipment, including a V-2 combustion chamber, to take back to Russia. As for the British, their experts were allowed into the area, but only after the Soviet team had done its work. When a larger group of experts back in the Soviet Union got their first look at the recovered parts, they were shocked by their findings. As Russian rocket pioneer Boris Chertok recalled: "We had at the time liquid engines for our experimental rocket planes with thrusts of hundreds of kilograms. One and a half tons was the limit of our dreams. Yet here we quickly calculated, based on the nozzle dimensions, that the engine thrust was at least 20 tons." Another time, he said: "In many respects for our future [rocket] activities, Churchill's appeals to Stalin were truly decisive."

The Soviet team also recovered parts from the V-1 cruise missile. The aviation officials who were in charge of the operation decided they were more interested in building craft like the V-1, which was similar to jet aircraft of the time, than the V-2. The work of reproducing the V-1 went to a group headed by Vladimir N. Chelomei, who became one of Russia's top rocket and spacecraft designers and a major rival to Korolev and his successors. Like Korolev, Chelomei was born in Ukraine and studied at the Kiev Aviation Institute. Chelomei's efforts to reproduce the V-1 had limited success but laid the foundations for the work of his design bureau.

Chertok was an aircraft engineer who had worked on small Soviet jet aircraft and the BI rocket aircraft efforts during the war. The sight of the V-2 parts made him look forward to getting his hands on the remains of the German rocket program. But the Soviet armed forces weren't interested in rockets, certainly not the same way they were interested in the German jet aircraft or in nuclear weapons. Despite this, Chertok and others were able to persuade some influential people that the rocket parts and factories in Germany might be valuable. When Soviet forces entered Peenemünde in May 1945, they found it deserted and most of the rocket material gone. Similarly, when the Americans handed them the Mittelwerk plant that summer, they found that most of the rockets and tooling had been spirited away by the American forces. And the leading figures of the German rocket group, they learned, were negotiating to go to the United States. "What could be more revolting and more inexcusable?" Stalin raged. "How and why was this allowed to happen?" Like the Americans and others in Germany at the time, the Soviets' initial efforts to grab the fruits of the German rocket efforts were disorganized, but that began to change once the war ended.

Inside Germany, a growing number of Soviet experts were scrutinizing the remaining V-2 parts, including some secreted out of a mine and others found abandoned in rail cars, to learn what secrets they contained. That summer, Chertok took a group of experts—most of whom had been sent to Germany as newly commissioned military officers—and established them in Bleicherode, a town not far from the Mittelwerk plant. Glushko and other Soviets moved into the Villa Franka, where von Braun had lived for a time late in the war, and then went looking for their own German rocket experts. They quickly found low-level engineers and set them to work with Soviets in Bleicherode in what they called the Institute RABE, from the German words for missile construction and development. Some of the Russians and Germans began test-firing V-2 engines on test stands at Lehesten, also near the Mittelwerk plant. Chertok also set up what he called Operation Ost (German for East) to persuade German experts to work for the Russians. Many spouses and children of German rocket experts were in the Bleicherode area, and some of the Russians established a rapport with the women. One of the women told them that Irmgard Gröttrup, the wife of physicist Helmut Gröttrup, who had helped lead development of the V-2's flight control system and was arrested by the SS in 1944 along with Wernher von Braun, wanted to meet with them. Chertok said she told him that she "hated fascism" and that she too had been arrested by the Nazis. Bargaining hard with the Russians, she won her husband and his colleagues the chance to continue their rocket work with good salaries and food rations at a work site inside Germany. With the move of the Gröttrups to the Russian institute at Bleicherode, the Russians had made their biggest catch, almost under the noses of the Americans. The Russians wanted more experts, however, including von Braun, and one Russian officer even went to the American zone and tried unsuccessfully to persuade an American officer that the German experts were war booty and should be shared.

As the Russian government struggled in the latter part of 1945 to shift from war with Germany to its new and uneasy peace with the West the ultimate disposition of the Russian rocket effort was still very much at play. Chertok and his associates often had to explain their work to conflicting and sometimes unfriendly authorities in their own military. Maj. Gen. Lev Gaydukov had become convinced of the importance of rockets, so he personally intervened with Stalin to ensure that rocket research was kept in one place and not split up among many ministries or brought under the control of the secret police. After the ministries responsible for aircraft production and ammunition production rejected the rocket work, Dmitry Ustinov, who was in charge of armament and artillery production for the Red Army, decided to take up the missile research. "In the initial period of Soviet missile technology development," Chertok wrote, "we were fortunate to have devotees and individuals with initiative and daring." The Russian aircraft experts sent to Germany had converted themselves into missile enthusiasts: "They refused to follow the orders from their immediate and very highest authorities to curtail their activity in Germany," Chertok recalled. "Finally, after the usual postwar delays, we had found a strong boss [Ustinov] in the industry for the new technology."

That fall, new experts came to Germany from Russia, thanks in part to Gaydukov's intervention, among them Korolev and Glushko. Korolev clearly reveled in his freedom after years as a prisoner of the secret police, a freedom enhanced by the luxurious conditions the conquering Soviet officers enjoyed in Germany. His most treasured perk was being issued his own car. Korolev, who was unknown to many of his new colleagues in Germany because of his former status as an enemy of the people, set up his own research group and joined the Soviet delegation that witnessed the launch in October of a V-2 at Cuxhaven on the North Sea by Germans under British supervision. Korolev received special unwanted attention from American officers, who forced him to watch the launch from a distance.

Early in 1946, Korolev returned to Germany from a short trip to Moscow with a promotion to colonel, which put him on a par with other Russian experts in the military pecking order. In March, Gaydukov announced the formation of the Institute Nordhausen, which consolidated all the Russian rocket research underway in Germany, including Korolev and Glushko's work in Berlin, Chertok's guidance group at the Institute RABE, Gröttrup's laboratory, and other facilities. Gaydukov was put in charge of the new institute, but significantly, Korolev, who just a few weeks before had ranked below most other experts, was now Gaydukov's deputy.

The Russians continued to learn as much as they could about the captured German technology and began assembling about a dozen V-2s from leftover parts. Although the Soviet nuclear weapon and aircraft programs received top priority, on May 13, the Soviet Council of Ministers passed a decree setting up a structure for missile development, which included a governmental committee similar to the one running the nuclear weapons program. The committee was headed by Georgi Malenkov, who after Stalin's death would serve for a short time as Soviet premier, with Ustinov as one of the deputy chairs. In accordance with this decree, which was the organizational charter for the Soviet missile efforts, Ustinov set up Scientific Research Institute No. 88, designated NII-88 after its Russian acronym. He also appointed Maj. Gen. Lev Gonor as first director and Korolev as chief designer of long-range missiles. The positive impression Korolev had made on Gaydukov and Ustinov protected him against those who opposed the appointment of a former political prisoner to such a sensitive post. Korolev in turn appointed as his deputy Vasili Mishin, a young engineer who had worked with Chertok and others on rocket planes during the war. Mishin would serve in this capacity for the rest of Korolev's life. The new institute was established at a plant in the northern suburbs of Moscow, in an area once known as Podlipki, then Kaliningrad, and today Korolev.

In the fall of 1946, preparations began to transfer the rocket work back to the Soviet Union. The Russian officers were asked to decide which Germans they wanted to continue work in Russia. On the night of October 21, the Germans were invited to a party at a local restaurant. They were allowed as much food and drink as they wanted, while the sullen Russians, who were prohibited from drinking that night, looked on. At 4 a.m. the next morning, the German scientists were awoken by Soviet security police and bundled along with their families and furniture onto trains for Russia. About seven thousand experts in rocket, aircraft, and nuclear projects and their families were forcibly deported to Russia, including about five hundred Germans involved in the rocket program.

By January 1947 the operations at Institute Nordhausen had been wound down. Chertok later said that the work in Germany gave the Soviets a chance to study at first hand the strengths and weaknesses of the German rocket technology. By starting in Germany, the work began with the benefit of Germany's advanced technology, and it also attracted the attention and support of the Soviet leadership. Chertok also quoted Korolev: "The most valuable thing we achieved there [in Germany] was forming the basis of a solid creative team of like-minded individuals." In this respect, the formation of this team of Russian rocket experts gave them an important leg up on the Americans, who took longer to exploit von Braun and his group as well as their captured rockets.

As for the German rocket engineers, once they settled in their new homes in the Moscow area, they worked alongside Russian experts in their plants. But Korolev and Glushko were unenthusiastic about continuing to work with the Germans, and while the Russians tried to learn as much as they could from the Germans, they ensured that the foreigners were excluded from important tasks. One factor that ensured the Germans' continued employment was Stalin's penchant for duplicating Western equipment. This was most famously expressed in the Soviet Tupolev 4 bomber, which was reverse-engineered from American B-29 Superfortresses held by Soviet authorities after they landed in the Soviet Union in 1945. Stalin desperately wanted a nuclear weapon to match the bombs the Americans had dropped on Hiroshima and Nagasaki, and he also wanted the means to deliver those weapons. Under the direction of Igor Kurchatov, Soviet scientists were put to work building nuclear weapons. In the case of rockets, Stalin wanted a Soviet copy of the V-2, which developed by Soviet engineers with the Germans' help under Korolev's direction became known as the R-1.

The Germans took part in launches of V-2s at the new Soviet launch site at Kapustin Yar, near Stalingrad, which began in October 1947. Despite its semiarid nature, the new launch location reminded Irmgard Gröttrup of "Peenemünde when we made our first experiments."

The Germans also designed a new rocket that offered several advances over the V-2, but their ideas were rebuffed by the Russians. Early in 1948, most of the Germans were moved to a scenic but isolated facility on Gorodomlya Island in Lake Seligar in Russia's upper Volga region, where they worked on their own on new designs and on small tasks for the Russians. By 1950, Gröttrup was so frustrated that he resigned, and most of the Germans were returned to East Germany by the end of 1953. Gröttrup moved to West Germany and spent the rest of his career in the electronics industry after rebuffing offers to return to the aeronautics industry.

Although the German rocketeers who went to Russia did not make the long-term contributions to the space program there that the von Braun team did in the United States, Asif Siddiqi argues in his history of the Soviet space program that the Germans' work constituted "an essential catalyst" for Soviet rocket work. "Without the help of the Germans, the Soviets—and in particular NII-88—would have clearly lagged in their efforts." The Soviets were far behind the Germans in 1945, and Gröttrup and his group transferred the German advances to the Soviets, greatly hastening their progress in rocketry.

Postwar Soviet Rockets

Even with the help of the Germans, it's impossible to overestimate the achievement of the Soviet rocket engineers in building the world's most advanced rockets right after a war that had killed millions of Soviet citizens and caused untold hardship to those who survived. In the immediate postwar years, Soviet food supplies were very poor, and disease was a major problem. Korolev's biographer James Harford records that Korolev had to spend a great deal of time in those years simply looking after the personal welfare of those who worked for him.

After the remaining V-2s were launched, many of them carrying scientific instruments, the plains at Kapustin Yar trembled in 1948 to the similar sounds of the R-1s, which began flying successfully after some early failures. Korolev, however, was more interested in working on the R-2, a larger version of the R-1 powered by engines designed by Glushko's bureau. Developing this new rocket had been very difficult, not least because many Soviet industries such as advanced metallurgy were backward or nonexistent. When the R-2s began flying in 1950, every single one failed, yet the rocket still represented an important advance for Korolev and his growing team.

About this time, several scientific research institutes were set up to deal with specific rocketry issues. After the heads of these groups held infor-

15. This photo of the Soviet Council of Chief Designers dates from 1959. From the left are Aleksei Bogomolov, Mikhail Ryazansky, Nikolai Pilyugin, Sergei Korolev, Valentin Glushko, Vladimir Barmin, and Viktor Kuznetsov. When Korolev formed the council in November 1947, it operated as an informal and separate entity from the other institutes and design bureaus and eventually assumed engineering control over much of the early development of the Soviet space program. (NASA photo)

mal meetings, Korolev formed a Council of Chief Designers that included himself and other leaders such as Glushko, guidance chief designer Viktor Kuznetsov, launch facility chief designer Vladimir Barmin, radio systems chief designer Mikhail Ryazansky, and Nikolai Pilyugin, chief designer for autonomous control systems. Since the chief designers worked in different ministries, the council gave missile design and production a needed forum for coordinating work. In 1945, Mikhail Tikhonravov, a GIRD veteran who survived the purges and the war, began work on ideas for crewed spacecraft and the multistage rockets needed to get humans into space. Although Korolev demonstrated his great interest in these ideas when he attended a talk by Tikhonravov in 1948, Tikhonravov's effort was suspended from fear that the secret police would look askance at any work that did not focus on the Soviet Union's security needs. In 1950, several sectors in NII-88 were merged into the Special Design Bureau No. 1, known as OKB-1, headed by Korolev. Gonor had left NII-88 in 1948, and when his successor, Konstantin Rudnev, was promoted in 1952, many people were surprised to see Mikhail K. Yangel, one of Korolev's subordinates, win promotion over Korolev for NII-88's top job. Yangel had a long record of activity in the Communist Party, and as later events would show, he was Korolev's peer as a leader in the missile industry. Korolev did little to conceal his unhappiness about the decision. The following year, Korolev was finally given the top job, along with membership in the Communist Party and in the Soviet Academy of Sciences. Yangel was given his own research institute in Dnepropetrovsk in Ukraine, where his team developed intercontinental ballistic missiles (ICBMs) for the Soviet military.

Before the R-2 flew, Korolev began work on a larger and more advanced rocket known as the R-3. Though the R-3 was to have a range of eighteen hundred miles, making it the first Soviet ICBM, it was held back by several problems, notably Glushko's struggle to build an engine that burned kerosene and liquid oxygen in place of the alcohol and liquid oxygen combinations of the V-2 and its Soviet successors. The R-3 experience caused Korolev and his team to step back and devise a rocket that, if less ambitious than the R-3, still represented progress. So in 1951, they began work on the R-5, a rocket that contained many of the innovations first earmarked for the R-3 but with a range of only 750 miles. After some initial problems, the R-5 flew a successful test series from Kapustin Yar during 1953. The success demonstrated the Soviet rocket team's growing confidence and opened the door for work on an ICBM—a vehicle that could deliver a nuclear warhead to the United States and, incidentally, launch payloads into space.

From very early on, the Soviet rockets had carried scientific payloads. In the late 1940s, Korolev began thinking of launching living beings into space in preparation for human flights one day. Together with aviation medicine specialists, Korolev's engineers deliberated in 1950 whether to launch apes or dogs. They chose dogs because Russian scientists had more experience with canines and believed they would be better behaved in flight. (For their animal flights, the Americans chose primates because they closely resemble humans.) To fly the dogs, the Soviets prepared a capsule that would separate from a modified R-1 rocket known as the R-1V, which was also fitted with experiments loaded into two modules that rode on the rocket's sides. On July 22, 1951, two dogs were chosen from a pool of nine at Kapustin Yar and put aboard the capsule. Dezik and Tsygan (Gypsy) were launched to an altitude of sixty-three miles and reached a speed of twenty-six hundred miles per hour. They experienced four minutes of weightlessness and parachuted safely to earth. They were the first living beings to survive a trip into space, faring better than the unfortunate primates that had flown earlier in Amer-

16. Sergei Korolev, founder of the Soviet space program, in July 1954 with a dog that had just returned to earth after being lobbed to an altitude of one hundred kilometers on an R-1D scientific rocket. (NASA photo)

ica's Project Blossom. During the next few weeks, five more rockets were launched with dogs. Two flights failed, killing four of the dogs, including Dezik on her second flight. Dezik's loss so grieved Korolev and the others that Tsygan never flew again and became a pampered pet. Overall, the animal program was considered successful, leaving a set of measurements of the animals' reactions to the stresses of flight and successfully testing life-support and recovery systems for the dogs' capsules.

Stalin's death in March 1953 just before the R-5 first flew apparently deeply affected Korolev, who like many who suffered during the dictator's three decades of rule, blamed others for his imprisonment. In Stalin's final months, a new round of purges had begun, this one with an anti-Semitic element. Several people in the Soviet apparatus faced persecution simply for being Jews, including Boris Chertok, who was demoted. Many were afraid of what might come after Stalin, and the leadership that succeeded him included Lavrenti Beria, the dreaded secret police chief who had a major say in military and scientific programs, along with Georgi Malenkov, Nikolai

Bulganin, and Nikita Khrushchev. The others soon had Beria arrested and executed, and eventually Khrushchev emerged as the new leader.

Thus, at Stalin's death, Korolev had many reasons to look to the future with apprehension. Stalin had personally made the big decisions about Soviet rocketry, and the others in the leadership had little knowledge of the subject, even Malenkov, who had nominal responsibility for rocketry in the immediate postwar years. Khrushchev later vividly captured the state of the new leadership's knowledge of rocketry in his description of their first meeting with Korolev:

I don't want to exaggerate, but I'd say we gawked at what he showed us as if we were a bunch of sheep seeing a new gate for the first time. When he showed us one of his rockets, we thought it looked like nothing but a huge cigar-shaped tube, and we didn't believe it could fly. Korolev took us on a tour of the launching pad and tried to explain to us how the rocket worked. We were like peasants in a market place. We walked around the rocket, touching it, tapping it to see if it was sturdy enough—we did everything but lick it to see how it tasted.

Khrushchev and his colleagues left many of the important decisions in the hands of people like Ustinov, and the missile programs continued. Indeed, Korolev's final ascent to the leading position within NII-88 and the Soviet rocket and space programs took place in the months following Stalin's death. In 1957, the year after Khrushchev lifted the veil on Stalin's crimes, the government notified Korolev that he was rehabilitated in the eyes of the Soviet state and acknowledged that he had been unjustly imprisoned.

The R-7

In the immediate postwar years, the Soviet nuclear program was given top priority. Scientists under the leadership of Igor Kurchatov, with the assistance of Soviet agents in the West, made impressive progress, shaking the West when they successfully exploded an atomic bomb on August 29, 1949. The Soviets exploded their first hydrogen bomb on August 12, 1953, at Semipalatinsk, less than a year after the first American H-bomb explosion. With the successful development of these weapons, the pressure on Korolev's team to develop a delivery vehicle intensified. Physicist Andrei Sakharov, one of the key developers of the Soviet H-bomb, was called upon to make an estimate that would shape the upcoming space race between the Soviet Union and the United States. The Kremlin wanted to be able to launch H-bombs to the United States aboard ICBMs, so Vyesheslav Malyshev, then minister of Medium Machine Building, which ran the missile program, asked Sakharov to estimate the weight of the H-bomb that an ICBM would have to carry. Sakharov's reply—five-and-a-half tons—was based on a concept for the final weapon that was later abandoned. Nevertheless, Malyshev ordered Korolev and his team to develop an ICBM that could carry a warhead of that weight, though it was nearly twice the three-ton weight the rocket team had thought they needed. Thus, in the fall of 1953, Korolev and his team began work on the first Soviet ICBM, the R-7, a rocket that, due to Sakharov's estimate, would be substantially larger than the first American ICBMs.

Glushko and his group of engineers continued to struggle building more powerful engines up to the task of lifting the payload decreed by the government. A way around the heating and burning stability problems these engines encountered was to build engines with multiple chambers instead of one. The concept, which involved a single turbopump feeding four thrust chambers, led to the four-chambered RD-107 and RD-108 engines on the R-7 and solved several technical problems. The rocket's basic design also represented an innovation. Instead of having a second stage attached to the top of the first stage, the R-7 consisted of a hammerhead-shaped core vehicle eighty-five feet high that was surrounded by four conical strap-on rockets, each sixty-two feet long. The four strap-ons constituted the first stage, and all four separated simultaneously while the core or second stage continued to burn.

In the mid-1950s, Korolev's and Glushko's teams worked hard on these and other challenges, notably accurately guiding a missile to a target thousands of miles away. The R-7, which became known as the Semyorka or "Old Number Seven," required a new launching system to support the ungainly vehicle. Since the existing launch site at Kapustin Yar was believed to be too close to American listening posts in Turkey, it was thus decided that a whole new launch site was needed. In 1955, the Soviet government chose a site in a semiarid region in Kazakhstan on the railroad line between Moscow and Tashkent near a small village called Tyura Tam. Construction of the first R-7 launch pad began that year. When the man placed in charge of the new facility, Maj. Gen. Alexei Nesterenko, arrived there by plane in the summer, he compared his first steps to a walk "into a blazing furnace." That wasn't all. "When we arrived at the construction site, our hearts sank: there was nothing but naked steppe, not a tree in sight, with only piles of sand and an assortment of animals scattered across the countryside." As construction continued into the winter, the workers had to contend with bitterly cold weather exacerbated by the winds that had free rein in the area. Over two years, the workers built a giant launch pad that stood over a huge pit carved from the desert. Nearby they built an enormous assembly building for the R-7 rockets and facilities to house the equipment and people who would run the launch site.

While the large industrial effort required to build the R-7 gained momentum, Korolev's missiles still had to prove themselves to a most important client: the Soviet military. The R-5 rocket would be useful to the military because it could carry a nuclear warhead to most targets in Europe. Tests of a military version of the R-5 went ahead in 1954 and 1955, and on February 2, 1956, an R-5 rose from Kapustin Yar bearing a live nuclear warhead. The flight, known as Baikal, went off without a problem, and the warhead exploded as planned at a designated impact site. "For Korolev and his engineers at NII-88," historian Siddiqi had noted, "this was a watershed moment. The years of uncertainty and suspicion from military leaders evaporated in a flush of euphoria." Soon top government officials visited NII-88, and the rocketeers received medals for their work. With the R-5 soon added to the Soviet military arsenal as the first nuclear-armed missile, the effort to build the R-7 ICBM won top priority.

While Korolev, Tikhonravov, and others concentrated on the R-7, they remained aware that it was more than a missile that could loft nuclear warheads but also a launch vehicle that could put satellites into orbit around the earth. With the paranoia of the Stalin era subsiding, in 1953 Tikhonravov won official approval to begin studies of satellites, and he soon put his group to work on the practical problems of making them a reality. The following year, his group completed an important report laying out ideas for satellites that could send scientific information back to earth. The report also discussed sending humans into space and returning them safely home and sending spacecraft to the moon.

In 1950, scientists from several countries began preparing for an eighteen-month period of coordinated international research set for 1957 and 1958 that would be known as the International Geophysical Year (IGY). In 1954, the United States proposed that satellites be launched as part of the IGY, and on July 29, 1955, the White House announced that the United States would do exactly that. About the same time, Leonid Sedov, the head of the Soviet delegation to the International Astronautical Congress in Copenhagen, called the press and announced that the Soviets too would launch a satellite during the IGY. Siddiqi has noted that Tikhonravov's 1954 satellite proposal had languished in government offices until after the White House announcement, when Korolev won approval from both the military and the Academy of Sciences to proceed with work on a satellite. When Soviet leaders visited NII-88 following the success of the R-5 nuclear launch, Korolev took advantage of their visit to lobby for the satellite. By then, a scientific commission headed by Mstislav Keldysh had reported on scientific tasks for a satellite, and a few days later detailed work began on what was known as Object D, the satellite that would one day follow dummy nuclear warheads as a payload for the R-7. As work on the satellite and its booster went on during 1956, Object D fell behind schedule, however. Tests suggested that the R-7 engines would be slightly less powerful than hoped. Early in 1957, Korolev and Tikhonravov therefore agreed to start work on a new and much smaller satellite known as the Simple Satellite, while work continued on Object D.

Korolev had hoped to launch the first R-7 in early 1957 and a satellite before the IGY got under way on July 1, 1957, but the inevitable problems of such large programs slowed progress. Once static tests cleared the missile for flight, the first R-7 was raised on the pad at Tyura Tam on the afternoon of May 6 after a full dress rehearsal with the rocket a few days earlier. When the launch took place on the evening of May 15, a strap-on rocket failed ninety-eight seconds into the flight, causing the rocket to disintegrate. Heat from nearby engines was a major cause, so a second rocket with improved heat shields at its base was installed at the pad. Three attempts to launch the rocket ended in aborts, and so the rocket was carried back to the assembly building. The third rocket was launched on July 12 but failed just thirty-three seconds after liftoff. Though greatly concerned by the failures, Korolev and his team tried again on August 21. This time the R-7 performed flawlessly, delivering its dummy warhead to an impact zone four thousand miles east in Kamchatka. The failure of the warhead's reentry vehicle to prevent the payload from disintegrating above the tar-

17. This 1959 photo is of the so-called "Three Ks" of the Soviet military-industrial complex. From far left are Sergei Korolev from the missile program, Igor Kurchatov from the atomic bomb program, and Mstislav Keldysh from the Academy of Sciences. On the extreme right is Korolev's first deputy, Vasili Mishin. (NASA photo)

get was the only problem encountered, but this was eclipsed by the rocket's success. Six days later, the Soviet media announced that the Soviet Union had successfully launched an ICBM that opened the door "to launching missiles into any region of the terrestrial globe." The Western media gave little prominence to the Soviet announcement, perhaps because so little was known about Soviet progress with rockets, perhaps because little credence was given to news originating from the Soviet government. For his part, Korolev spent little time savoring his success; he was planning another launch that would spark a different reaction.

8. The Military-Industrial Complex

We have the scientists and the engineers. We enjoy industrial
superiority. We have the inventive genius. Why, therefore, have
we not embarked on a major space program equivalent to that
which was undertaken in developing the atomic bomb?

The editors of Collier's *magazine, March 22, 1952*

As he lay on his deathbed in August 1945, Robert Goddard read newspaper
stories announcing the dropping of two atomic bombs on Japan. Though
Goddard knew that the destruction of Hiroshima and Nagasaki meant
World War II was over, we can only speculate whether he realized that
the creation of atomic bombs would in a few years' time give new life to
the rocket technology he had worked so hard to foster. In any case, Goddard's death marked the end of the age of single inventors in the field of
rocketry and space exploration. The Germans had assembled their rocket
team before the war began, and members of that group were transferred
to the United States shortly after the war ended. The nucleus of the Soviet
Union's rocket team had been formed in the 1930s. Those who survived Stalin's purges and his erratic wartime leadership started again, almost from
scratch, after 1945. The exigencies of war had shown the American military, industrial, and academic establishments that large groups were essential to building the weapons needed to win the war. The best example of
this was the atomic bomb itself, designed and built by a team of American
and British scientists under the supervision of the U.S. Army Corps of Engineers in the Manhattan Project.

In comparison to Europe, the war had barely touched America. But the
casualties of the far-off battlefields and the rationing at home were real
enough to most Americans. Moreover, the war had followed immediately

after the Depression, which had caused hardship for so many people. So at war's end, the U.S. government concentrated on bringing its soldiers home and shifting its economy from military needs to providing the luxuries that Americans craved. America had defeated its enemies, and Russia had suffered greatly in the war. The United States alone possessed nuclear weapons and the bombers that could deliver them almost anywhere on earth. What's more, the U.S. Army had scooped up Germany's top rocket engineers and most of their V-2 rockets, denying them to the Soviets. And where brute force wasn't appropriate, hopes were high that the newly founded United Nations would prove more effective than the League of Nations, which had failed in the 1930s. In this atmosphere, there was little appetite for spending money on new weapons such as rockets. And until there were rockets, space exploration remained in the realm of fantasy. Skepticism about the capability of rockets remained alive and well, even in the best-informed circles. In December 1945, the U.S. military's top scientist, Dr. Vannevar Bush, told a U.S. Senate committee what he thought about intercontinental missiles: "In my opinion such a thing is impossible and will be impossible for many years."

Though Goddard's work inspired many Americans to think about going into space using rockets, his refusal to work with others allowed Werner von Braun's Germans to move ahead of him even before war broke out. Two groups of Americans, one on the east coast made up of members of the American Rocket Society, and another on the west coast, formed the foundation of America's postwar progress in rockets.

Caltech and von Kármán

The west coast group was built at the California Institute of Technology or Caltech, which was based in Pasadena, a wealthy suburb of Los Angeles located at the foot of Mount Wilson in the San Gabriel Mountains. In the 1920s and 1930s, Caltech's leaders worked to attract the world's top scientific talent. Albert Einstein himself spent time at Caltech before settling in Princeton. Robert Millikan, the first American-born scientist to win a Nobel Prize, established Caltech's physics department and then turned to developing expertise in the field of aeronautics. Aviation was becoming big business in the Los Angeles area, with firms such as Douglas Aircraft, Lockheed, North American, Consolidated Vultee, and Hughes already building aircraft there. In 1929, Millikan's efforts paid off when the distinguished Hungarian physicist and engineer Theodore von Kármán agreed to come to Caltech.

Von Kármán, one of the most brilliant scientists of the twentieth century, used mathematics to explain the complex problems of aerodynamics. He was known for his flamboyant personality and his inattention to the mundane details of life. He lived most of his life with his mother and a devoted sister, and although he was known to pursue attractive women, he never married. Born in 1881 in Budapest, then one of Europe's cultural centers, von Kármán was the son of a distinguished educator and a mother descended from a family of scholars. His family benefited from the tolerance shown to Jews at the time. He first exercised his talent for mathematics in the high school his father founded, which also turned out other famous scientists such as John von Neumann, Leo Szilard, and Edward Teller. Von Kármán capped his undergraduate work in engineering at the Royal Joseph University of Polytechnics and Economics with a landmark 1906 paper on the buckling of structures. That year he left Hungary for Germany, where he studied engineering at the University of Göttingen, then one of the world's top scientific universities. He chafed under the rigid hierarchies he found at Göttingen, however, so in 1908, von Kármán went to Paris to consider his future. There, he saw his first aircraft in action, barely four years after the Wright Brothers made their first flight.

The sight of the aircraft left von Kármán fascinated by the problems of aviation, and he returned to Göttingen to pursue studies in this new field. By 1913 he escaped Göttingen's hierarchies by accepting a professorship at the Technical University in Aachen and the directorship of the Aachen Aerodynamics Institute. World War I interrupted his academic career, however, and he served during the war as an engineer in the Hungarian army. Von Kármán was finally able to return to Budapest at war's end, and there he was persuaded to serve as undersecretary of education in the short-lived communist government of Béla Kun. That was von Kármán's last venture in politics. He later said his time in government "saved me for all my life from having any belief in Communism—I saw it in operation and that was sufficient." He soon returned to Aachen and rebuilt his institute, which had been closed during the war. With the help of his mother and sister, von Kármán established himself in comfort near the university, and built his worldwide reputation with a series of studies in aerodynamics. Much of his work was supported by aircraft manufacturers such as Junkers and Zeppelin.

By the mid-1920s, however, the hyperinflation that struck Germany had reduced enrollment in von Kármán's courses and along with it financial support for his work. Such economic dislocations, coupled with discontent arising from Germany's loss in the war, fanned anti-Semitism and the rise of Adolf Hitler's Nazi Party. Von Kármán was denied a prestigious post at Göttingen because he was a Jew. By then, Millikan and Caltech had begun courting him. In December 1929, the Hungarian scientist left Germany and came to Pasadena. He and his academic output flourished in Caltech's informal setting, and von Kármán established Caltech as a force in the field of aeronautics. With financial help from the Guggenheim family, which was also backing Goddard's rocket research, the university established the Guggenheim Aeronautical Laboratory at the California Institute of Technology (GALCIT). Through this organization, von Kármán conducted his work and probably more importantly, trained graduate students who would play pivotal roles in bringing America into the supersonic age and into space, including William Bollay, Frank Malina, Tsien Hsueshen, and Apollo M. O. Smith.

Malina was a young Texan who possessed an interest in the arts and strong liberal opinions in politics. His scientific work was motivated by his dreams of spaceflight, which originated with boyhood readings of Jules Verne. In 1936, he began a rocket research project along with Bollay, Smith, Tsien and two other rocket enthusiasts: John W. Parsons, a chemist, and Edward S. Forman, a mechanic. They began testing rocket motors in the desolate Arroyo Seco, a dried-out riverbed canyon near a dam on the edge of Pasadena, just a short walk from where Robert Goddard and Clarence Hickman had tested solid rockets during World War I.

The following year, Malina's rocket group won von Kármán's and GALCIT's support, including use of GALCIT's building on campus. The rocket tests were moved outside when one test of a tiny motor left a cloud of methyl alcohol and nitrogen dioxide around the building, depositing a layer of rust on most of the equipment. The group also acquired a name they are still remembered by: the "suicide squad." Although its membership changed over time, the squad continued both theoretical work at Caltech and tests in the Arroyo Seco on a liquid-rocket engine, which led to more positive notices. Of course, their rocket tests resulted in the inevitable failures and explosions so familiar to anyone who tried to advance the science of rocketry. Their liquid-fuel work included rocket engines that used gasoline and red fuming nitric acid as an oxidizer. This new oxidizer was poisonous and corrosive, but it also didn't need to be handled at low temperatures like liquid oxygen. The value of teamwork and the availability of brilliant theoreticians such as Tsien and von Kármán helped the testing along. During this period, both von Kármán and Malina tried and failed to convince Goddard to work together with them.

In 1938, with war clouds beginning to gather, Gen. Henry H. "Hap" Arnold, the farsighted head of the Army Air Corps, as it was called before 1941, visited Caltech. Soon Malina was invited to speak before a scientific committee supervising Air Corps research whose membership included von Kármán. The committee funded five research projects, including one given to GALCIT to work on rocket-assisted takeoff units for heavy aircraft. Aeronautical engineer Jerome C. Hunsaker of the Massachusetts Institute of Technology agreed to run a research program on deicing windshields and turned down the rocket work. "You can have the Buck Rogers job," he told von Kármán.

"The remark made by Hunsaker reflected the general attitude that prevailed in engineering circles as regards rockets and rocket propulsion," Malina wrote years later. "The word 'rocket' was of such bad repute that von Kármán and I felt it advisable to drop the use of the word. It did not return to our vocabulary until years later." Instead, they used the word *jet*, which is why rockets used to help aircraft take off are still called "jet-assisted takeoff units" or JATO, and why the name GALCIT adopted during the war—the Jet Propulsion Laboratory—didn't contain the more accurate word, *rocket*.

Armed with a $10,000 contract in 1939, GALCIT set to work on a number of liquid-fueled and solid-fueled rocket units to help aircraft take off. The liquid rockets were difficult to handle because of the oxidizer and, in some cases, because of the corrosive fuels used. Solid-fuel rockets were easier to handle, but no solid rockets had yet been developed that fired longer than three seconds—not long enough to help aircraft take off from short runways. As well, solid rockets tended to explode after even short periods of storage. Eventually, theoretical work by von Kármán and engineering development work by Malina and his colleagues led to a workable solid-

18. Dr. Theodore von Kármán (black coat) sketches out a plan on the wing of an airplane as his Jet-Assisted Take-Off (JATO) engineering team looks on. From left to right: Dr. Clark B. Millikan, Dr. Martin Summerfield, Dr. Theodore von Kármán, Dr. Frank J. Malina, and the pilot, Capt. Homer Boushey. Captain Boushey would become the first American to pilot an airplane that used JATO solid-propellant rockets. (NASA photo)

fueled JATO unit for aircraft in 1941. Air Corps captain Homer A. Boushey, himself an engineer and a friend of both the GALCIT researchers and Robert Goddard, made the first successful flight using JATO units in 1941. Later that year, Boushey made the first American flight of a rocket-powered aircraft, the Ercoupe, which used twelve JATO units. In 1942, the GALCIT research overcame the JATO storage problem. These new JATOs, which were fueled by asphalt and other easily obtained materials, were quickly picked up by the navy to facilitate takeoffs from aircraft carriers. They also formed the foundation for further advances that would establish the United States as the leader in solid rocketry in the 1950s and 1960s.

Robert Goddard also worked for both the U.S. Navy and the Army Air Force during the war developing liquid-fueled JATO units. His efforts proved to be of limited use to the armed forces, however, because the GALCIT units using solid fuels proved to be safer and easier to handle.

The GALCIT group's success with JATO and the U.S. entrance into the war in 1941 led to more and larger military contracts to develop new rockets and related devices for the military. Malina, von Kármán, and other members of the suicide squad formed a company to build and sell JATO units to the military. The company, known as the Aerojet Engineering Corporation and later as Aerojet General, also kept the word *rocket* out of its name, even when it became one of the top companies in the field of rocketry.

In 1943, von Kármán began receiving intelligence reports about the German rocket effort at Peenemünde. The news of this German rocket caused the U.S. Army Ordnance Corps to ask GALCIT, newly renamed Jet Propulsion Laboratory or JPL, to begin work on long-range missiles. Under the direction of von Kármán and Malina, JPL began to grow as it developed solid-fueled "Private" rockets and liquid-fueled "Corporal" rockets. With military funding growing into the millions of dollars, JPL established itself in buildings on the once deserted Arroyo Seco and extended its research load into theoretical work on supersonic flight and jet engines.

In late 1944, with work on the Private and Corporal rockets well underway and the end of the war in sight, Malina persuaded the army to support development of a scaled-down version of the Corporal rocket for use as a sounding rocket that could carry twenty-five pounds of equipment to an altitude of at least one hundred thousand feet. Because the rocket was considered a "little sister" of the Corporal, the rocket was called "WAC Corporal" for the Women's Army Corps. Some soon noted that WAC could also be taken to mean "without attitude control."

By the fall of 1945, the WAC Corporal began flying in tests at White Sands, New Mexico, setting an altitude record exceeding 235,000 feet. The rocket flew with scientific payloads on suborbital trajectories before the German expatriates could get their V-2s at White Sands flying. In 1949, the WAC Corporal was married to the V-2 in Project Bumper. These two-stage vehicles were the first objects made by humans to break into space. The Johns Hopkins University Applied Physics Laboratory later collaborated with Aerojet to build a larger version of the WAC Corporal, the Aerobee, which became the United States's workhorse sounding rocket for many years. More than one thousand Aerobees lofted scientific instruments to high altitudes until it was retired in 1985.

The growth of JPL into a major laboratory doing millions of dollars of work for the military inevitably meant changes. Parsons, the brilliant chemist who had been involved in a black magic cult, left JPL in 1944 and Aerojet the next year. He died in an explosion in 1952 in his home laboratory. Von Kármán effectively left JPL at war's end to do government work in Washington. Malina, who was happier working on research rockets such as the

WAC Corporal than on military programs, left JPL in 1947 to take a post with the United Nations Education, Scientific and Cultural Organization (UNESCO). By then, the shadow of the Cold War was not only stimulating military research but also fostering intolerance of those who held left-wing opinions. Discussion groups from the 1930s such as one that Malina and Tsien had joined came under the FBI's scrutiny because communists had also taken part in the groups.

The most notorious security case involving JPL and Caltech centered on Tsien. He was visited by the FBI, and then in 1950 he had his security clearance taken away, a few months after Mao Zedong's communists had seized power in China and after Tsien had decided to apply for U.S. citizenship. Tsien generally avoided politics, but his participation in the discussion group was enough to make him a target of the FBI. An angry Tsien decided to return to China, but he soon found himself in a bureaucratic limbo—he was under a deportation order but couldn't leave the United States. He was also jailed for a time. Eventually, in 1955, Tsien and his family were permitted to leave for China, where Tsien became the leading light of China's missile and space programs. China launched its own satellite for the first time in 1970, and Tsien lived to see China's first "taikonaut" launched into orbit in 2003. Other scientists also were caught up in the anticommunist repression in the United States. Von Kármán, for example, was called in for questioning about his activities in Béla Kun's government in 1919, and J. Robert Oppenheimer, the scientific director of the Manhattan Project, lost his security clearance in 1953 because of his political leanings.

Aside from the development of the WAC Corporal, most of the effort at JPL in the first decade after the war focused on tactical rockets for the U.S. Army. The first was the liquid-fueled Corporal, deployed with army units, which was limited because of the problems inherent in liquid fuels, including the need to fuel them just before launch. It became clear that solid-fuel rockets would be more useful to the army, and so JPL proceeded with the solid-fuel Sergeant missile, which reached maturation in the 1960s. In the late 1940s, JPL continued its work advancing solid rockets. JPL's asphalt JATO rockets still had many drawbacks, and a JPL engineer found that a new polysulfide developed by the Thiokol Chemical Corporation had the potential to improve on asphalt's burning and handling characteristics. JPL engineers found that with a star-shaped inner hole running the length of the solid propellant, the rocket burned at a slower and more controllable rate. Shortly after this breakthrough, JPL wound down its solid rocket research. Nevertheless, its findings had laid the foundation for gigantic solid-fuel rockets that would be used as ICBMs and submarine-launched ballistic missiles, and later to boost spacecraft into orbit. Another legacy of JPL's military work in the 1940s and 1950s was the large number of people who went on to help design and build U.S. missiles and launch vehicles. Louis G. Dunn, for example, who succeeded von Kármán as director of JPL at war's end, left the lab in 1954 to work in the Atlas missile program. Because JPL's expertise extended to guidance and communications systems for rockets, this expertise came to the fore later in the 1950s when it began looking for new initiatives beyond military rocket programs.

Elsewhere at White Sands in the late 1940s, the U.S. Navy also pursued its own rocket program, Viking, which was designed to provide an alternative to the V-2 as a launcher for larger scientific payloads. The V-2 had many limitations, including the fact that its nose cone had to be fully loaded for the rocket to be fully stable and that it could not be controlled once its engine cut off at only twenty miles altitude. The size and power of the Viking was a third that of the V-2, but it had more sophisticated control systems that made it a superior platform for scientific experiments. The navy gave the Viking contract to the Martin Company, an aircraft contractor looking for work after the war, and its engine was made by Reaction Motors Inc., which had been set up by members of the American Rocket Society. The engine designer, John Shesta, drew on his experience in society activities and also on information gleaned from the V-2's rocket engine.

Starting with a first flight from White Sands in May 1949, Viking had a largely successful run, although along the way the Viking team learned the usual hard lessons about the difficulties of rockets. On one occasion, a rocket tore loose from its restraints during a static test and was destroyed after less than a minute of flight. When the firing button was pushed to launch *Viking 10*, an explosion destroyed the engine and drained the oxygen tank, leaving the rocket standing amidst a conflagration as its alcohol fuel drained slowly. Someone in the blockhouse concluded that the upper part of the rocket could be saved if the sealed alcohol tank could be opened before it collapsed on itself. The solution was to shoot a hole in the tank atop the rocket. As dimples began to appear in the tank from the falling pressure inside, Viking director Milton Rosen authorized the shot from a carbine, which relieved the falling pressure in the tank and allowed *Viking 10* to be repaired and flown another day. There were also happier moments. *Viking 4* soared more than 100 miles high after being fired off a navy ship in the Pacific, following an earlier shipboard firing of a V-2. *Viking 7* broke the V-2's single-stage rocket altitude record when it flew 135 miles high, and later more powerful Vikings flew even higher.

The Germans Move to Huntsville

In 1949, the Germans at Fort Bliss had nearly used up their V-2s. They also needed to relocate because the army authorities in El Paso had decided to return the base hospital annex, which had served as their home, to its original use. The Germans were on their way to becoming U.S. citizens, and so Holger Toftoy, by then an army major general, began looking for a new home for his rocket experts. He found it in Huntsville, Alabama. This northern Alabama town's economy had relied on cotton until World War II, when the army established the Huntsville Arsenal to manufacture poison gas, which was never used in the war, and the Redstone Arsenal, where artillery shells were made. The end of the war meant that most production ceased, and Huntsville's economic good times came to an end. At first, Toftoy's plans to move to Huntsville drew little interest because the town hoped to attract an air force engineering facility, but when the air force chose another site, Toftoy quickly won local support to move the army's rocket program there. Not everyone in the army was keen on rockets, and Toftoy, in one memorable pitch to army leaders, had crawled on the floor over maps he'd spread there to point out the attributes of Huntsville's arsenals and facilities. "I'm really on my hands and knees, literally and figuratively begging for this place," he'd implored. "Are there any questions?" The deepening Cold War, which the successful Soviet detonation of an atomic bomb and the communist takeover of China had intensified, helped the army leadership decide in favor of Toftoy's ambitious rocket plans.

Between April and November 1950, the German group, then standing at 130 due to new arrivals and some departures, moved to Huntsville along with more than 500 military personnel, 120 civilian employees, and hundreds of employees of General Electric, which was contracted to help the army rocket program. While the Germans moved to their new homes in

Alabama, the invasion of South Korea by troops from communist North Korea marked the beginning of the Korean War. America responded with a large commitment of troops and forces. Some of the Americans who flew jets in combat in the skies over Korea would eventually populate places like Edwards Air Force Base and Pax River. For the German rocket team, the Korean War meant that the army was at last interested in using their talents to develop new rockets. The Germans and General Electric had been working on a rocket development program called Hermes, but the lack of army support produced little tangible progress. In 1950, the army issued specifications for a missile that could carry a three-ton warhead two hundred miles, and the team at the Ordnance Guided Missile Center in Huntsville got to work on the Redstone missile, named after the arsenal. The Germans were aided by the Chrysler Corporation, which helped assemble the missiles, and the Rocketdyne division of North American Aviation, which built the missile's engine. The first sixteen Redstones were built at the arsenal in Huntsville, after which Chrysler took over manufacturing. When the first Redstone was launched on August 20, 1953, it failed, as did the second, but with the third flight success came to the program. Though the Redstone has often been compared to the V-2, it contained multiple innovations, particularly in its guidance and control systems, many the work of the Germans during their time in America.

The arrival of the Germans in Huntsville marked a new phase in their integration into American society, a phase that was capped on April 15, 1955, when twelve hundred people turned out to witness von Braun and forty other German experts, along with their families, become U.S. citizens. Five years earlier, when a Huntsvillian had seen a moving van draw up to von Braun's new home, he thought to himself, "Here goes the neighborhood to the dogs." But soon the man and his family became close friends with the von Brauns. The Germans set up a Lutheran church in the community, although von Braun himself chose to worship at the local Episcopal church. The Germans became keen members of local service clubs, and one became president of the Junior Chamber of Commerce even before gaining his citizenship. Von Braun's parents lived temporarily with him in Huntsville before they returned to Germany after gaining pension rights from the postwar German government, and Baroness von Braun kept busy teaching languages. The Germans established a symphony orchestra and an astronomy club in Huntsville, along with the facilities required for both. The new Americans built homes and saw to it that local libraries and schools were of the highest standard. The once sleepy cotton town of Huntsville became Rocket City U.S.A. and was never the same again.

With the Redstone missile under way and his family growing with the birth of two daughters, who were later joined by a son, von Braun was able to dedicate some of his time to an activity he would prove particularly skillful at: promoting space exploration. On Columbus Day 1951, a symposium on space travel took place at the Hayden Planetarium in New York. About two hundred experts participated, including Willy Ley, who had become a popular writer on the subject in America since leaving Germany in 1935, and British writer and theorist Arthur C. Clarke, though not von Braun himself. Two writers from the mass-circulation magazine *Collier's* were there, and they convinced their editors that space travel was a subject worthy of serious attention. One of their editors, Cornelius Ryan, who later gained fame for his best-selling book on the 1944 Normandy invasion, *The Longest Day,* was skeptical. Nevertheless, he went to a space medicine meeting in San Antonio that was addressed by von Braun. Ryan was converted by von Braun's arguments, and he began assembling a series for the magazine that featured von Braun and many of the experts who spoke at the Hayden Planetarium symposium.

The series began in the March 22, 1952, issue of *Collier's*, with a readership of millions of Americans. "Man Will Conquer Space Soon. Top Scientists Tell How in 15 Startling Pages," the cover proclaimed above a dramatic Chesley Bonestell painting of a rocket at the moment of staging. The articles included warnings that America must gain control of space before the Soviet Union. "Within the next 10 or 15 years, the earth will have a new companion in the skies," von Braun warned, "a satellite that could either be the greatest force for peace ever devised or one of the most terrible weapons of war—depending on who makes and controls it." It went on to describe a space station that could be used for scientific purposes—and to give its operators an important strategic advantage in any nuclear confrontation. Over the next two years, seven more issues of *Collier's* contained articles on space exploration, including concepts for rockets, satellites, space vehicles, and human flights to the moon and Mars. The articles and illustrations were also sold in book form. To promote the *Collier's* series, von Braun gave interviews on radio and a new medium, television.

19. Dr. Wernher von Braun (right), then chief of the Guided Missile Development Operation Division at the Army Ballistic Missile Agency (ABMA) in Huntsville, Alabama, was visited by Walt Disney (left) in 1954. In the 1950s, von Braun worked with Disney Studio as a technical director, making three films about space exploration for television. (NASA photo)

As the *Collier's* series was winding up, Ley and von Braun were brought in contact with someone who would give them their biggest and best opportunity yet to sell their ideas—Walt Disney. In 1954, Disney had already made a fortune from cartoons featuring his stable of characters headed by Mickey Mouse and from feature-length movies, many of them animated, aimed at family audiences. That year, Disney was building what he called a leisure park and today is known as a theme park. Because his plans for Disneyland represented a new concept, Disney had difficulty attracting investors. Disney's solution was to sign a deal with the American Broadcasting Corporation, the smallest of the three U.S. television networks, for investment in his park in exchange for a series of hour-long TV broadcasts that promoted the new park and showed Disney features created specially for the show. *Disneyland* premiered in the fall of 1954, a few months before the park of the same name opened, and the show became a Sunday night institution during the early years of television. Both the park and the show contained four theme areas: Fantasyland, Frontierland, Adventureland, and Tomorrowland. The futuristic last area included a simulated ride around the moon, and Disney tapped one of his top creative people, Ward Kimball, to produce shows on the Tomorrowland theme. Since fantasy already had its own category, the Tomorrowland shows had to be factual, and Kimball recruited Willy Ley to begin work on shows on space travel. When the first show, "Man in Space," appeared on TV on March 9, 1955, more than forty million Americans were watching. After introductions by Disney and Kimball, Ley, von Braun, and space-medicine pioneer Dr. Heinz Haber gave viewers a crash course in the basics of space travel, with a major assist from humorous but educational animation produced by Kimball. The show wrapped up with an animated depiction of von Braun's idea of a space flight that resembled the space shuttle flights of the late twentieth century. Late that year, the second show, "Man and the Moon," appeared. Following an animated history of human ideas about the moon, the show featured a talk by von Braun and a dramatic simulation by live actors of a flight around the moon, complete with an earthrise. Two years later, the third and last show, "Mars and Beyond," showed an animated flight to Mars using atomic-powered spacecraft. All three shows appeared more than once on television, and they were also used as short subjects in movie theatres. Space exploration was now on the American agenda, and von Braun was becoming a celebrity.

Air Force Missiles

During the postwar years, the branches of the U.S. armed forces jockeyed for control of America's U.S. missile forces. The army worked with JPL and the Germans, and the navy backed Viking. The U.S. Air Force, even before its establishment in 1947, also claimed jurisdiction over missiles. Hap Arnold departed from the air force just before it became a separate service, and one of his parting gifts was a massive report from a group headed by von Kármán, titled *Toward New Horizons*, which foresaw supersonic jet aircraft and ICBMs in the air force of the future. In 1946, the air force signed contracts with a number of aircraft firms to work on missiles that would pay dividends down the road. In return for its proposal for a long-range missile, the Consolidated Vultee Aircraft Company in San Diego, better known as Convair, received a $1.9 million study contract from the air force for a program known as MX-774. A group at Convair under the leadership of Belgian-born Karel "Charlie" Bossart then began work on a missile that would eventually be known as Atlas. Although Convair got a taste of the challenges of developing missiles with MX-774, the money soon ran out and so did air force interest in missiles. Three MX-774 rockets were launched at White Sands before the air force money ran out, and Convair continued work on a considerably reduced basis using company funds.

Other firms were looking at pilotless aircraft, today known as cruise missiles, with Northrop building a missile known as Snark and Martin another known as Matador. Most importantly, North American Aviation won a contract to work on a more complex cruise missile, the Navaho. At the end of World War II, North American was looking for new work as aircraft orders dried up. The future lay in jet aircraft, and the firm hired William Bollay, an alumnus of von Kármán's group at Caltech. During the war, Bollay, who had been born in Germany but came to the United States in 1924 when his family emigrated, applied his expertise to jet and rocket engines in the U.S. Navy. North American put Bollay to work on the Navaho, which was planned to be an intercontinental cruise missile. The Navaho had a booster rocket with two and later three rocket engines that carried a cruise missile powered by two ramjets to altitude. The Navaho concept proved too complex for the time, however, and it was cancelled in 1957 when ballistic missiles were overtaking it. Bollay had left North American long before then, but while there he developed the rocket engines for the Navaho's booster. At first Bollay envisioned a winged version of the V-2. Since more power was needed than the V-2 engines could deliver, he developed an engine that generated 75,000 pounds of thrust. This engine formed the basis of the Rocketdyne division of North American and was quickly picked up by von Braun for the Redstone. Soon the Navaho grew and needed more powerful engines of 120,000 pounds of thrust. Newer versions of the Rocketdyne engines were powered up even more and used for the Atlas and other ICBMs. Although only a few Navaho missiles flew before the program was canceled, its engines were used to power America's early ICBMs and satellite launch vehicles.

The explosion of the Soviet atomic bomb in 1949 and the start of the Korean War caused the air force to look again at ICBMs, and in 1951 Convair won a new contract to resume its work on ballistic missiles. Bossart and his team had concluded that an ICBM would require a two-stage rocket, but they were concerned about starting the second-stage engines in flight. They decided to avoid this problem with an innovative stage-and-a-half design in which all engines would light at liftoff, and sustainer engines would burn throughout flight while the other engines were dropped off. Another innovation that saved weight and dated from the MX-774 was its metallic skin, which was thinner than a dime. This thin skin required that the rocket's tanks be pressurized with gas when it was not fueled in order to prevent the tanks from crumpling, but these pressurized tanks, known to many as "steel balloons," were sufficient to carry the rocket's aerodynamic loads in flight.

Early in its development, the missile, to be known as Atlas, was a five-engine behemoth because most experts thought that was what would be needed to carry nuclear warheads. But while the development work went on, both the United States and the Soviets exploded hydrogen bombs. In 1953, Gen. Dwight D. Eisenhower became president, and his administration adopted a new policy for defense known as the New Look, which stressed the use of cost-effective strategic nuclear weapons over expensive conventional forces. The Russian H-bomb made the matter of ICBMs more important to the U.S. military, so the Air Force Science Advisory Board appointed a committee headed by the Hungarian-born mathematician John von Neumann to look at nuclear weapons and delivery methods. Not only did the committee recommend proceeding with ICBMs; it found that powerful H-bombs with an explosive force of half a megaton—roughly four hundred times as powerful as the bomb that leveled Hiroshima—would soon weigh as little as 1.5 tons. Further research soon revised this weight downward to three-quarters of a ton. Because the primitive guidance systems of the time limited missiles' ability to accurately strike targets thousands of miles away, these more powerful bombs were needed to ensure that a target would be destroyed. The announcement that lightweight nuclear weapons were feasible opened the door to ICBMs that were smaller than originally estimated, and therefore less complex and expensive.

The new administration in Washington also brought a champion of ICBMs into the Pentagon in the person of Trevor Gardner, who became the assistant secretary of the air force responsible for research and development. Gardner and air force secretary Harold Talbott knew they had a fight on their hands with generals determined to protect their bombers during a time of budget retrenchment. Gardner tapped von Neumann again to head a special Strategic Missiles Evaluation Group, nicknamed the Teapot Committee, to look at existing missile programs with the help of America's top scientific talent and researchers from RAND. "The aim was to create a document so hot and of such eminence that no one could pooh-pooh it," Gardner later said. As Gardner hoped, the Teapot Committee concluded in February 1954 that America's missile programs were in disarray and that the United States could not assume that the Soviet Union was not making progress with its own ICBMs. Convair's Atlas ICBM needed to be given top priority. The complexity of building the ICBM mandated the creation of a special agency to develop and manage the missile program so that, in the committee's words, "over-all technical direction [will] be in the hands of an unusually competent group of scientists and engineers capable of making system analyses, supervising the research phase, and completely controlling the experimental and research phases of the program."

The air force acted quickly to implement this recommendation, establishing the Western Development Division in the suburbs of Los Angeles to manage ICBM programs. The division was placed under the command of Gen. Bernard Schriever. The newly formed Ramo-Wooldridge Corporation, later TRW Inc., was hired to provide independent systems engineering and technical advice as Convair set to work on the Atlas, now reduced from five engines to three thanks to the smaller warheads it would have to carry.

Schriever proved to be the crucial figure in driving America's ICBM programs and its first generation of launch vehicles. This was primarily because of his espousal of management methods such as systems management and configuration control, both of which were needed to run complex technical programs. Schriever also took care to ensure that he had the authority he needed to build the ICBMs as he saw fit. He left a legacy not only in the rockets and spacecraft that the Western Development Division and its successor organizations built, but in a group of managers inside and outside the military that guided American missile and space programs for years into the future.

One of Schriever's management methods was to ensure that more than one approach and one contractor were responsible for major subsystems. In fact, this dual-track approach was taken so far that a second ICBM system, the Titan, was built by a different contractor, Martin, and with different design features and subcontractors. The division was also assigned responsibility for an Air Force Intermediate Range Ballistic Missile (IRBM), the Thor. So as the United States moved into the second half of the 1950s, Schriever and his air force contractors were building Atlas, Titan, and Thor, while von Braun and his army group were flying the Redstone and building the Jupiter, another IRBM.

The teams headed by von Braun and Schriever were not only building various vehicles for delivering America's nuclear weapons; they were constructing the infrastructure of what would become America's space program. While these missiles were being built, America's leaders also had to make decisions about what they might carry into space.

9. Sputniks and Muttniks

Considerable prestige and psychological benefits will accrue
to the nation which first is successful in launching a satellite.
The inference of such a demonstration of advanced technology and its unmistakable relationship to inter-continental ballistic missile technology might have important repercussions
on the political determination of free world countries to resist
communist threats, especially if the USSR were to be the first
to establish a satellite.

National Security Council NSC-5520, May 20, 1955

Arguably the most famous dinner party in the annals of twentieth-century
science took place on April 5, 1950, at the home of physicist James A. Van
Allen in the Washington DC suburb of Silver Spring, Maryland. The guest
of honor was the distinguished British geophysicist, Sydney Chapman. Another guest was Lloyd V. Berkner, a radio engineer who shifted into scientific research after serving on the first Antarctic expedition of Adm. Richard Byrd. Also present were three other geophysicists—J. Wallace Joyce,
S. Fred Singer, and Ernest H. Vestine. All of these scientists had taken
part in the Second International Polar Year, a worldwide geophysical research effort in 1933. So when conversation turned to the state of geophys-
ics, Berkner suggested that advances in science recommended the obser-
vance of a third polar year in 1957 and 1958, a period of maximum sunspot
activity. The idea won the support of everyone present, and after international scientific bodies approved the concept, scientists from around the
world began work on what became known as the International Geophysical Year (IGY), a worldwide series of scientific observations to be conducted
in every part of the earth and its atmosphere. The IGY took place from July

1, 1957, to the end of 1958, and ultimately involved sixty thousand scientists from sixty-six countries.

During planning for the IGY in 1954, the idea was raised of launching a satellite containing scientific instruments to measure the upper atmosphere and space from earth orbit, complementing the IGY's other activities. The idea won approval from scientists like Van Allen, who was already launching "rockoons"—small instrumented sounding rockets launched from platforms lifted by balloons to high altitudes—to probe the upper atmosphere, and Singer, who in 1953 proposed a research satellite called MOUSE, or Minimum Orbital Unmanned Satellite, Earth, to transmit data about earth's magnetic field and conditions in space from an orbit two hundred miles above earth. In the fall of 1954, with strong urging from U.S. scientists, the IGY's coordinating committee passed a resolution calling for the launch of a scientific satellite during the IGY.

Scientists were far from the only ones thinking about satellites in the early 1950s. The U.S. Air Force and the U.S. Navy were battling for jurisdiction over activities beyond the atmosphere as they contended for control of missiles. At war's end, the navy's Bureau of Aeronautics supported studies into the High Altitude Test Vehicle, a rocket powered by liquid hydrogen that it hoped could loft satellites into orbit with a single stage. Because budgets were tight, the navy asked the Army Air Force in March 1946 whether it would be interested in joining in an in-depth study of satellites. The responsible air force official, Maj. Gen. Curtis LeMay, turned the navy down and then commissioned a satellite study from Douglas Aircraft. The air force wanted the study done quickly so it could establish primacy in the field of satellites. Douglas set up the "Research And Development" project, or Project RAND, which completed a 336-page report in May called *Preliminary Design of an Experimental World-Circling Spaceship*. The classified report contained engineering information on rockets that was based on the Germans' V-2 data and discussed what kinds of rockets were needed to loft satellites into orbit. The RAND report's introduction famously predicted that satellites could become "one of the most potent scientific tools of the Twentieth Century" and that launching a satellite "would inflame the imaginations of mankind, and would probably produce repercussions in the world comparable to the explosion of the atomic bomb. . . . To visualize the impact on the world, one can imagine the consternation and admiration that would be felt here if the United States were to discover suddenly that some other nation had already put up a successful satellite." A chapter in the report by Dr. Louis Ridenour noted that the requirements for an ICBM are similar to those for a satellite launcher, and he highlighted the idea, first raised early in the century, that satellites would be useful to the military for reconnaissance.

Project RAND—which soon split from Douglas and became the RAND Corporation, still a prominent think tank today—also conducted a more detailed technical study of satellites in 1947. After a pause, the air force initiated more satellite studies using RAND experts, including a study of satellites for military observation. A RAND study conference in 1949 examined satellites as a political and psychological weapon, specifically the impact that the existence of U.S. satellites would have on the leaders of the Soviet Union. And while the air force fended off occasional navy attempts to take over satellite programs, it commissioned RAND to carry out more detailed development work on reconnaissance satellites in the early 1950s under Project Feed Back.

Aside from the scientists of the IGY, others inside and outside the military were also talking about satellites. In 1945, the British space visionary and novelist Arthur C. Clarke published articles in a radio enthusiasts' magazine called *Wireless World* proposing that satellites be orbited at twenty-three thousand miles above the equator, where they would remain stationary relative to the earth. Today, such geosynchronous satellites form a vital part of the world's communications links. Von Braun and his collaborators in the *Collier's* series wrote about satellites and space stations, and in 1954, von Braun held discussions with officials from the Office of Naval Research about launching a small satellite atop a Jupiter C rocket, an uprated version of the Redstone.

The administration of President Dwight Eisenhower set up the Technological Capabilities Panel (TCP) in 1954 as part of a series of reviews of American defense policy. The TCP gathered several top U.S. scientists under the chairmanship of MIT president James Killian to consider measures to prevent surprise attacks on the United States similar to the Japanese attack on Pearl Harbor in 1941. This meant overcoming the Soviet secrecy that the American military found so frustrating, especially after the successful Soviet nuclear bomb tests starting in 1949. The United States tried a number

20. *Sputnik 2* is shown at the launch pad at Tyura Tam on November 3, 1957. Warm air was piped into the capsule to keep the dog Laika, the first living being to reach orbit, comfortable amid the freezing temperatures. (NASA photo)

of ways to penetrate Soviet secrecy, including placing radio and listening posts close to Soviet borders, flying aircraft near and sometimes into Soviet airspace, and even equipping balloons with cameras (an idea that not only failed to yield valuable information but gave the United States a diplomatic black eye). In 1955, Eisenhower proposed an "Open Skies" policy to the Soviets that would have permitted mutual inspection of military installations. The Soviet Union rejected the idea out of hand.

When the TCP issued its report early in 1955, its numerous recommendations included the creation of a reconnaissance aircraft that became the U-2 and large reconnaissance satellites capable of photographing aircraft and missiles on the ground. The TCP also called for a small artificial earth satellite to establish "freedom of space" for future U.S. reconnaissance satellites. Scientific satellites could obtain important data for military uses and reinforce America's reputation for technological superiority. Nine weeks after the TCP report, the U.S. National Security Council approved a document known as NSC-5520 that endorsed the TCP's recommendation for a small satellite because it would be a "small technological step" toward a reconnaissance satellite and would test the principle of freedom of space. "However," the council observed, "preliminary studies indicate that there is no obstacle under international law to the launching of such a satellite." NSC-5520 also recommended that the satellite program be associated with the IGY. There is evidence that Berkner, who was one of the top organizers of the IGY, was fully aware of the security implications of orbiting a satellite as outlined in NSC-5520. This fact suggests that Berkner worked with the U.S. military in promoting a peaceful scientific satellite for the IGY that would establish the right of overflight for U.S. satellites.

In a well-coordinated announcement flowing from the findings of NSC-5520, White House press secretary James C. Hagerty announced on July 29, 1955, that the United States planned to launch small satellites as part of the IGY, which led to a Soviet response that it would launch a satellite of its own "in the near future."

The satellite plan that von Braun had discussed with the Naval Research Laboratory in 1954 resulted in competing proposals from the laboratory to build a satellite named Vanguard and a competing proposal from the U.S. Army Ordnance Corps for a satellite named Orbiter. In August 1955, a committee headed by Jet Propulsion Laboratory scientist Homer Joe Stewart chose the navy satellite. It was a controversial decision since many have suggested that the weaker Vanguard proposal was chosen only because the Eisenhower administration was anxious to establish the right of overflight, or perhaps disinclined to have the first U.S. satellite sit atop a rocket built by Wernher von Braun's former German rocketeers. In his recent examination of the Stewart committee's decision, historian Michael Neufeld has argued that Vanguard was superior in many ways to the Orbiter proposal and that Orbiter was quite different from the Explorer satellites that von Braun's team subsequently launched during the IGY. For example, Orbiter was to be a five-pound satellite with no instruments, whereas Vanguard would be instrumented and carry a tracking system. He also found no evidence that the Stewart committee considered the issue of satellite overflight. Moreover, the Stewart committee's decision was upheld by higher authorities, and the U.S. government made Vanguard its IGY satellite program.

Vanguard, which used a three-stage rocket based on the Viking research rocket, fell behind schedule, however, and soon vastly exceeded its original budget as its engineers and scientists grappled with the complexities of

21. Gen. John B. Medaris (left), who was the commander of the Army Ballistic Missile Agency (ABMA) at the Redstone Arsenal in Huntsville, Alabama, from 1955 to 1958, shakes hands with Maj. Gen. Holger Toftoy (right), who consolidated U.S. missile and rocketry development after World War II. (NASA photo)

building a vehicle capable of launching satellites. The program was also troubled by the fact that many of the most qualified engineers in the United States were either in von Braun's army team or were building the Titan and Atlas ICBMs for the air force. The contractor's attention focused on the Titan, which was being built by the same contractor that was fabricating Viking's first stage. By the time the IGY began in the summer of 1957, the first Vanguard satellite launch had been delayed to 1958. The program also had funding difficulties, which were eased by a financial contribution from the Central Intelligence Agency.

In 1956, von Braun's team became part of the newly established Army Ballistic Missile Agency (ABMA), which had an energetic and flamboyant new leader, Maj. Gen. John Medaris. Both von Braun and Medaris chafed under the restrictions that prevented them from using their Jupiter C vehicle to launch a satellite. Jupiter C, an upgraded Redstone fitted with upper stages made from small solid-fuel rockets, was used to test nose cones for ICBMs. A Jupiter C launch in September 1956 carried a dummy payload 682 miles high and reached three-quarters of the speed needed to get into orbit. Von Braun and Medaris later claimed that officials from Washington came to Cape Canaveral to personally inspect the vehicle to ensure that it would not "accidentally" go into orbit. The competition between the armed forces to control missiles hardly cooled off when in 1956 Secretary of Defense Charles E. Wilson decided to give long-range missiles to the air force while limiting the army to missiles flying less than two hundred miles and the navy to shipboard missiles. Wilson's action infuriated Medaris, who said morale in Huntsville "reached an all-time low."

Another U.S. satellite program got under way in 1956, known as Weapons System 117L, under an air force contract to Lockheed Aircraft. The program to build a U.S. military reconnaissance satellite evolved from the ideas laid out in the Project Feed Back reports completed by RAND in 1954, which envisioned a satellite that could transmit its photographs back to earth for immediate analysis.

Sputnik Is Launched

Even before the Soviets succeeded in getting their R-7 ICBM to fly on August 21, 1957, they began dropping tantalizing hints about their own satellite strategy. In June of that year, Soviet scientists sent a detailed memorandum to the IGY coordinating committee on their satellite plans, and in July, the Soviets published instructions on how to find their satellite in the skies after launch and giving its radio broadcast frequencies. Like the Soviet announcement of the successful ICBM flight, these announcements were not taken seriously by many people in the West. However, when a second R-7 flew successfully down the Russian ICBM test range on September 7, proving the R-7 as a weapon, the door had opened to using the next R-7 to launch a satellite.

Korolev had hoped to launch a satellite by September 17, the one hundredth birthday of Russian space pioneer Konstantin Tsiolkovsky. When that proved impossible, he announced at a meeting in Moscow that "in the nearest future the first test launches of artificial satellites of the earth with scientific goals will take place in the USSR and the USA." Before he left Moscow, Korolev met with other officials and set October 6 as the target date to launch the satellite, then known as PS-1, or "simple satellite 1." PS-1 was a polished sphere of aluminum alloy nearly twenty-three inches wide, sprouting four antennas and containing a radio transmitter, batteries, and an instrument that measured the temperature inside the satellite. The 184-pound satellite replaced a much larger and more heavily instrumented satellite known as Object D that had fallen behind in production months earlier.

Because scientists from around the world were meeting in Washington at the beginning of October to discuss rocket and satellite research for the IGY, Korolev feared that the United States would launch Vanguard or a satellite on von Braun's Jupiter C rocket before the scientists dispersed. So on the morning of October 3, the squat R-7 rocket with the satellite under its nose cone rumbled out of the assembly building at Tyura Tam and, with Korolev and other officials walking alongside, rode on its side on a rail car to the launch pad a mile and a half away. The next morning, Friday, October 4, fueling of the R-7 began, but the launch was delayed several times during the day. Twenty-eight minutes after midnight local time, the R-7 finally roared off the pad. Less than six minutes later, the satellite and its second-stage rocket were in orbit. A tracking station in Kamchatka picked up the satellite's signals, but Korolev put off any celebrations until the satellite passed over the launch site an hour and a half later and the radio picked up the beeps of PS-1.

When the orbit was verified, Korolev telephoned the Soviet leader, Nikita Khrushchev, who was visiting Kiev for meetings with Ukrainian officials. The news pleased Khrushchev, who shared it with others at the meeting. The next day, the official Soviet newspaper *Pravda* carried the long-winded and low-key announcement on its front page that the Soviet Union had launched an object into orbit: a sputnik, the Russian word for satellite.

Although many informed people in the scientific community knew that a Soviet satellite was imminent, the news was still a shock to most Americans, for whom the premiere of a new television series, *Leave It To Beaver*, was expected to be the evening's most memorable event. At about 6 p.m. that evening, October 4, a large group of scientists in Washington for an IGY meeting were attending a Soviet embassy reception, when a *New York Times* reporter present took a phone call from an editor informing him of

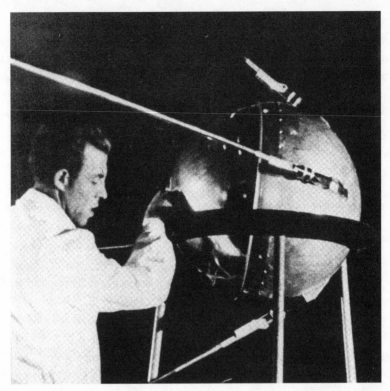

22. The Sputnik satellite is shown here on a rigging truck in the assembly shop in the fall of 1957 as a technician applies finishing touches. On October 4, 1957, Sputnik was successfully launched and entered Earth's orbit. The satellite shocked the world, giving the USSR the distinction of putting the first human-made object into orbit. (NASA photo)

Sputnik's launch. When he was told, Berkner called for everyone's attention and announced the news that turned the reception into a major celebration. News bulletins interrupted radio shows to broadcast the beeps of Sputnik, and years later, many people could still remember where they were when they heard the news. Ham radio operators scrambled to pick up Sputnik's signals, and others simply looked into the sky. Most people who did see a satellite actually saw the tumbling second stage of the R-7, not Sputnik itself. The next evening during the flight from Tyura Tam to Moscow, Korolev was told of the world's reaction to his feat. "Well, comrades," he told his colleagues, "you can't imagine—the whole world is talking about our satellite. It seems that we have caused quite a stir."

Before that fall Friday, most Americans didn't know what a satellite was or how it worked. But the fact that a craft launched by America's Soviet

Cold War adversary passed over U.S. territory several times a day was easy to understand, as was the fact that contrary to expectations, the Soviet Union had opened the space age rather than the United States. President Eisenhower reacted calmly and golfed on the weekend, but the following week, many Americans were showing concern and seeking answers. Von Braun, who was entertaining incoming Secretary of Defense Neil McElroy in Huntsville when the news of Sputnik arrived, immediately began lobbying to launch a satellite using his Jupiter C rocket.

The Russians, quickly sensing the unexpectedly strong world reaction to their achievement, began trumpeting Sputnik, giving it the headline treatment it had been denied the morning after the launch. To capitalize on the publicity, Khrushchev ordered Korolev to launch another satellite within a month to mark the celebrations of the fortieth anniversary of the Soviet Revolution on November 7. Object D was still far from complete, so Korolev and his team worked overtime to build a satellite from a duplicate of the original Sputnik but equipped with a cabin to carry a living passenger into space. *Sputnik 2*, with the dog Laika on board, was launched on November 3. The unfortunate mongrel died after only a few hours in orbit from excessive heating, though the Soviets tried to suggest she lived longer. The news that there was no way to return the dog to earth alive only slightly dampened the sensation created by the launch of the 1,118-pound payload. This time the satellite was even heavier because it remained attached to its R-7 second stage. The flight of *Sputnik 2* drove home two messages: the Russians could launch an atomic bomb that could hit any part of the United States, and they were working to launch humans into space.

Sputnik had clearly caught Eisenhower and his administration off guard. The president's attempts to downplay or rationalize Sputnik—including Chief of Staff Sherman Adams's dismissive reference to an "outer space basketball game"—fell flat. Political opponents such as Democratic Senate leader Lyndon B. Johnson took advantage of a rare opportunity to score political points against the popular president, launching widely publicized hearings on the state of America's rocket and satellite programs. Johnson and others called Sputnik a technological Pearl Harbor, and historian Walter A. McDougall has described the reaction to Sputnik as a "media riot." Four days after the launch of the second Sputnik, Eisenhower gave the first of a series of televised speeches from the Oval Office in the White House on the subject of "Science and Security." Next to his desk was a nose cone recovered from a Jupiter C launched earlier that year. The next day, the secretary of defense announced that von Braun's team had been given the go-ahead to launch a satellite atop a Jupiter C as a "supplement" to Vanguard, although the Huntsville team had already started to prepare a rocket for launch.

Vanguard and Explorer

At this moment, Vanguard was still proceeding according to plan. Two Vanguard test vehicles had flown before Sputnik went into orbit, and on October 23 a Vanguard rocket with a live first stage and dummy second and third stages flew as hoped from its Cape Canaveral launch pad. The next Vanguard launch, slated for December, was to test all three functional stages of the rocket. If the four-pound test satellite atop the rocket made it into orbit, it would be considered a "bonus." Sputnik magnified the pressure on the Vanguard team, and when the Vanguard TV-3 vehicle was launched on December 6, a national audience was watching on television. The rocket rose four feet, and then crumpled as the thrust ceased. Before a giant fireball enveloped the pad, the nose cone containing the satellite tipped off the rocket. America's humiliation was total, and headlines such as "Flopnik," "Kaputnik," and "Dudnik" bluntly described the nonlaunch. Despite later successes, Vanguard became a byword for failure. America's support for science and technology, its educational system, and even the viability of American capitalism began to be questioned from within. John Hagen, the director of Vanguard, charged that America had "gone soft" and urged more financial support for education and science. The Eisenhower administration came under renewed attack, as did the interservice rivalry that had contributed to America falling behind the Soviets in space. A few people pointed out that the secrecy of the Soviet program and the openness of America's efforts also contributed to the events of 1957.

While the recriminations sounded in Washington and around the country, the Vanguard team and von Braun's group raced to get an American satellite into space. The backup Vanguard launch vehicle and satellite to the ill-fated TV-3 rocket were erected on their pad. Four attempts to launch the rocket between January 23 and 26, 1958, ended in scrubs. Finally, when the second stage had to be dismantled to repair a damaged engine, the Vanguard team postponed their launch attempt until February.

23. Jet Propulsion Laboratory director Dr. William Pickering, Dr. James Van Allen of the State University of Iowa, and Army Ballistic Missile Agency technical director Dr. Wernher von Braun triumphantly display a model of the *Explorer 1*, America's first satellite, shortly after the satellite's launch on January 31, 1958. (NASA photo)

At a nearby launch pad, von Braun's army team erected a four-stage Jupiter C rocket with a thirty-one-pound bullet-shaped satellite on top. *Explorer 1* was built by engineers at the Jet Propulsion Laboratory in California. It carried a Geiger counter in an experiment package designed by James Van Allen to search for radiation in space, originally intended to fly on Vanguard. The army team had five days to launch before Vanguard would get another chance, but bad weather blocked launch attempts on January 29 and 30. Finally, late on the evening of January 31, the Jupiter C roared into the skies, this time without live television coverage. Like the Russians before them, the army team had to sweat out the satellite's first orbit before they knew if their launch had succeeded. The wait for confirmation of orbit was longer than expected because the Jupiter C had lofted *Explorer 1* into a higher orbit than planned. Finally, one hundred minutes after launch, a tracking station in Earthquake Valley, California, picked up the satellite. By then it was after midnight on February 1, but President Eisenhower, who had been playing bridge while awaiting confirmation of the satellite's fate, immediately announced the news on the radio. Von Braun, who had been forced to sweat out the launch in Washington DC rather than Cape Canaveral, joined Van Allen and JPL's director William Pickering for a jubilant news conference at the headquarters of the National Academy of Sciences. In Huntsville, Alabama, word of the successful launch kicked off an all-night celebration in the streets.

The news that America had launched its own satellite cooled American concern about Sputnik. Johnson, who had held widely publicized hearings in the weeks after Sputnik, "abandoned" his interest in the space issue, according to his biographer Robert A. Caro and had to be persuaded by aides to issue a report on his hearings. Though he turned to such issues as a recession and racial desegregation in the South, he and other politicians still kept an eye on space, as they demonstrated that year by working to give the United States a space agency and passing legislation to massively increase U.S. spending on education and scientific research.

Among its successes, *Explorer 1* had discovered radiation around the earth, though the data were limited because they came only in real time over a handful of tracking stations. But soon both the army and navy teams suffered new setbacks. The postponed Vanguard launch ended in failure a minute after launch on February 5, and *Explorer 2* failed to orbit in early March. Finally, the Vanguard team tasted success when *Vanguard 1* reached orbit on March 17, carrying a tiny four-pound satellite that was the first to use solar cells as a power source. *Explorer 3* was launched on March 26. Since *Sputnik 2*, the Soviet Union had been silent on the space front. The satellite that was originally to have been the first Soviet satellite, Object D, failed at launch in April. Its backup, which became *Sputnik 3*, weighed in at 2,926 pounds, loaded down with a full set of scientific equipment, including radiation detectors. It was launched on May 15, 1958.

The immense size of the third Sputnik revived American concerns about the Soviet ability to loft large payloads, but the Americans soon won a less visible race to obtain scientific results. Both *Sputnik 3* and *Explorer 3* carried radiation sensors, and both had tape recorders to comprehensively track radiation levels around the earth. The tape recorder on *Sputnik 3* failed, however, and as historian James J. Harford has written, the Soviets were "hogtied" without recorded data. Indeed, if *Sputnik 2* had carried a tape recorder or orbited within earth's radiation belts when it passed over the few Soviet tracking stations it might have been the first to report the unexpectedly high levels of radiation in space. After eliminating the possibility that the high radiation *Explorers 1* and *3* detected came from nuclear tests, Van Allen and his team deduced that the radiation lay in "belts" around the earth, and that summer they were officially associated with Van Allen's name. In July, *Explorer 4* probed further out into the Van Allen belts, as did three Pioneer probes launched later that year. Though the Americans had lost the race to get the first satellite into orbit, they beat the Russians into what was arguably the most important scientific discovery of the International Geophysical Year—the earth's Van Allen radiation belts.

Though a modest satellite with only the barest of instrumentation, *Vanguard 1* also made its contribution to science. It was sent into a relatively high orbit and observed by a specially established network of Minitrack stations. The data from the satellite's path showed that its low point dipped more when it was north of the equator than when it was south of it. The implication of this finding was that the Earth is slightly pear shaped—one of the more important discoveries of the IGY.

As both U.S. teams continued to launch satellites during 1958 and 1959, they found that the road to orbit is not an easy one. After *Vanguard 1*'s launch, the Vanguard team had to endure four failures in a row before *Vanguard 2* entered orbit on February 17, 1959, and began measuring the earth's cloud cover in a precursor mission to the weather satellites that would begin flying in the 1960s. After two more failures, the final Vanguard launch on September 18, 1959, carried *Vanguard 3* into orbit with a 51-pound package of scientific instruments. Meanwhile, after *Explorer 4*, the army team experienced a string of four launch failures before succeeding on October 13, 1959, with the Juno II rocket bearing the 91.5-pound *Explorer 7*, which also carried a number of scientific instruments.

Though *Explorer 1* is justly remembered as a great American technical triumph, those involved in Vanguard always believed that its own poor reputation was unjustified. "Vanguard started with virtually nothing in 1955, completed vehicle design in March 1956, and had a fully successful flight two years later," Vanguard's director, John P. Hagen, wrote. "One can challenge any other new rocket program in the United States to demonstrate a completely successful launching within such a short time." Von Braun, for his part, endorsed Hagen's belief in reflecting on his own team's success with *Explorer 1*: "It's not that we're geniuses. It is just that we have been working on these things so long we have already made more mistakes than the other people have."

Military Efforts

For many years, another U.S. military program to put a small satellite in orbit in the summer of 1958 was shrouded in secrecy. The program, the first attempt to launch a satellite from an aircraft, was officially named Project Pilot but has become known by the nickname of its satellite: NOTSNIK, named after the Naval Ordnance Test Station (NOTS) at China Lake, California, where the program was based. In the program a specially constructed booster fabricated from a number of small solid-fuel rockets was mounted on a Navy F-4D-1 Skyray aircraft based at NOTS. The small 2.3-pound NOTSNIK satellite was equipped with radiation detectors to gather information from nuclear tests occurring at the time. Six attempts were made in July and August 1958 to launch a NOTSNIK satellite into orbit, but most failed soon after launch. One disappeared shortly after launch, and though some speculated it might have made it briefly into orbit, no evidence supports this contention. These launches were not only the first to use a booster dropped from an aircraft but the first attempt to put a satellite into orbit using a solid rocket. Similar rockets built by the air force were tested in 1958 and 1959 as satellite interceptors, but the first successful satellite launch using a booster released from an aircraft didn't take place until 1990, when a Pegasus rocket dropped from a B-52 delivered Pegsat into orbit.

While the eyes of America were focused on the Vanguard and Jupiter C rockets at Cape Canaveral, Bernard Schriever's missile development program was moving swiftly toward construction of America's reply to Korolev's R-7. On January 25, 1957, the first Thor missile was launched, but it rose only a few inches before collapsing in a fireball. Though subsequent launches also failed, Thor finally flew successfully in September. Thor was deployed in Europe, and was soon made available to launch American space vehicles, becoming the basis of the Delta launch vehicle that was a workhorse rocket for NASA and commercial satellite launches for more than four decades. The first test of the Atlas ICBM took place in June 1957, but it failed a minute into its flight. Atlas faced a frustrating series of failures with its engines and guidance system that were eventually solved in November 1958. Although uncertainty still surrounded Atlas, the rocket was used to give the United States a boost in the space race on December 18, when an entire

Atlas missile fitted with 150 pounds of communications gear was launched into orbit. The launch, known as Project SCORE (Signal Communications by Orbiting Relay Equipment), was the first test of a communications satellite. A tape recorder on board picked up messages from the ground and relayed them back to earth. SCORE's first broadcast was a Christmas message from President Eisenhower. Because the SCORE satellite included the missile body, its total weight of 8,750 pounds was compared to the lighter weight of the Soviet Sputniks. But those comparisons ignored the heavy R-7 core stages that had also been lofted into orbit with the Sputniks. Like Thor, the Atlas was deployed for military purposes and became one of America's top space launch vehicles. Though many of the Atlas's more spectacular failures were immortalized in films that have been played repeatedly since, these early failures taught engineers important lessons about how rockets work.

The air force's missile work didn't stop with Atlas and Thor. On February 6, 1959, the first Titan missile was successfully launched. Subsequent launches would be needed to wring out Titan's inevitable problems, however. In the 1960s, Atlas and Titan ICBMs were deployed around the United States as part of America's new nuclear deterrent against a Soviet attack with nuclear arms. Both the Atlas and the Titan I had limitations because they needed to be fueled with kerosene and liquid oxygen, and the oxidizer's extremely cold temperatures required special handling. Once the rockets were fueled, they could remain ready for launch for only a short time. The Titan II ICBM marked a step forward because it used fuels that could be stored in the rocket and could be kept at alert status for days at a time. But there were limits to even the Titan II's availability because the fuels were highly corrosive, restricting the amount of time they could be kept in the missile. The solution was to build a solid-fueled ICBM that could be kept on alert status indefinitely. Only in the late 1940s had the group at JPL made sufficient progress for the use of solid fuels to become possible in larger rockets. When Schriever's team began work on ICBMs in 1955, he ordered more development work on solid fuels, and in 1958 the Minuteman program began. Solid-fueled Minuteman missiles began being deployed in 1963, soon replacing Atlas and Titan as America's major ICBM. Liquid-fuel rockets are more suitable for use as space launch vehicles because their fuel provides greater thrust, and so Atlas and Titan were used mainly for space exploration rather than as weapons. Thor, for its part, was soon replaced in its military deployment by solid rockets, but it remained in use to launch satellites as the Delta launch vehicle. The U.S. Navy began work in 1956 to build a solid-fueled missile, the Polaris, that could be launched from submarines.

The military development of ICBMs forms a crucial part of the space exploration story because military ICBM research led directly to the exploration of space, both in the United States and the Soviet Union. Had ICBMs been developed at another time when solid-fuel technologies were more advanced, the development of suitable satellite launch vehicles might have been delayed for many years. Fortunately, for the purposes of launching satellites, the liquid-fueled ICBMs were developed first. The evolution of the ICBM left its mark on the space race in another way. Because the Soviets were desperate to build an ICBM regardless of weight, the R-7 was a behemoth capable of carrying heavy warheads to America. For that same reason, it could launch large payloads into orbit. The American ICBM effort began behind the Soviets' pace because of America's early lead in bombers and nuclear weapons, but even when the American ICBMs were built, they could not carry the same payload as the R-7 because they were designed for lighter warheads. The primitive state of Soviet electronics and miniaturization translated into a lead in building heavier rockets and thus a lead in the early years of the space race with the United States. But the R-7's squat design made it even less suitable for use as an ICBM than the Atlas, and so Russian design bureaus scrambled in the late 1950s and early 1960s to match the American progress in ICBMs.

Amid the political distress that Sputnik caused Eisenhower, the president could take comfort in the fact that the flight of Sputnik had inadvertently but indisputably established the right to fly over foreign nations from space. As Assistant Secretary of Defense Donald Quarles told the president four days after Sputnik, "the Russians have in fact done us a good turn, unintentionally, in establishing the concept of freedom of international space." The United States needed a way to penetrate the secrecy of the Soviet Union, because at the time they only had occasional overflights by U-2 aircraft, which flew too high to be shot down but not high enough to be missed by Soviet radar. That would end in May 1960, when the Russians shot down a U-2 and captured its pilot, Francis Gary Powers. As Eisenhower faced the political fallout from Sputnik in the fall of 1957, the WS-117 reconnaissance satellite program was faltering from tight budgets and the fact that transmitting photos from space was proving to be a difficult proposition. Early in 1958, scientist James Killian and Edwin "Din" Land, who was known to Americans as the inventor of the Polaroid camera but was also a key figure in developing reconnaissance satellites, suggested to Eisenhower that the United States work to develop a reconnaissance satellite that sent back photographic film to earth in recoverable capsules as an "interim" program until photos could be transmitted back to earth. Thus, WS-117 was ended, and a new secret reconnaissance satellite program called CORONA began, while work on developing more advanced reconnaissance technologies continued.

Any launch into space in the late 1950s attracted great publicity, so the air force characterized early CORONA launches as scientific research satellites called Discoverer. Some "Discoverer" flights even carried mice to provide a plausible cover story. CORONA/Discoverer was designed to be launched from Vandenberg Air Force Base in California atop a Thor missile with a second-stage rocket called Agena. But the first Discoverer flight failed even before the launch when its Thor-Agena rocket exploded. Discoverer underwent a lengthy series of failures that ranged from explosions of the Thor or Agena to camera failures and botched recoveries of the film capsules. Finally, in August 1960, a capsule from *Discoverer 13* was successfully recovered from the Pacific. Because the capsule did not carry any photographic gear, it was put on show and the flag it carried was presented to President Eisenhower. The president was likely more gratified later in the month when *Discoverer 14*, carrying the full CORONA photographic system, took photographs of military installations in the Soviet Union and successfully returned them in a capsule that the air force plucked from the skies as it descended under its parachute. The pictures were of poor quality, but they still provided important information about the state of the Soviet military. Most importantly, the first and subsequent successful CORONAs showed Eisenhower and his successor John F. Kennedy that Khrushchev's boasts about turning out ICBMs "like sausages" were empty. The first launch of a satellite that would test the broadcasting of reconnaissance photos from space as originally envisioned for WS-117 took place in 1960. That same year also saw the first launch of a satellite with sensors to detect a nuclear attack on the United States. After overcoming their respective difficulties, these satellite tests led to viable systems that helped protect America and its allies in later years. The reconnaissance satellites that America and later the Soviets launched became a cornerstone of national security during the Cold War. Most of the story is still secret. The full story of CORONA only came out in 1995, and the generations of reconnaissance satellites that followed remain shrouded in secrecy. In 1967 President Lyndon Johnson said that the reconnaissance satellites had probably saved "ten times" what the United States was spending on all its space programs. "Because tonight we know how many missiles the enemy has," he admitted, "and it turned out our guesses were way off. We were doing things we didn't need to do. We were building things we didn't need to build. We were harboring fears we didn't need to harbor."

10. The Birth of NASA

The Roman Empire controlled the world because it could build roads. Later—when men moved to the sea—the British Empire was dominant because it had ships. In the air age, we were powerful because we had airplanes. Now the communists have established a foothold in outer space. It is not very reassuring to be told that next year we will put a better satellite into the air. Perhaps it will also have chrome trim and automatic windshield wipers.

Sen. Lyndon B. Johnson, 1957

Although Nikita Khrushchev's boastful dismissal of the American Vanguard satellite as a "grapefruit" played to American worries that Sputnik constituted a "technological Pearl Harbor," President Dwight Eisenhower sought to assure Americans that the Soviet lead in technology was more illusory than real. Eisenhower had access to classified information that he couldn't share with the public, yet he had far less information than he wanted about what the Soviets were doing. This information problem was just one of the challenges American policymakers faced as they decided how to run their space program in 1958. One of the many reasons Sputnik made it into orbit ahead of American satellites was that the Soviets operated in secret yet could keep track of the latest events in the open American program. The Soviet space program was at that time a grouping of often feuding design bureaus headed by chief designers—of whom Sergei Korolev was first among equals—that operated as a subsidiary of the Soviet military. When the race to the moon gained momentum in the mid-1960s, the structure of the Soviet space program was found wanting. But in the late 1950s, sheltered by secrecy, it was giving Khrushchev plenty to crow about.

In truth, much of the U.S. space effort also was carried on in secret, notably the CORONA reconnaissance satellite project. Yet the bulk of America's space efforts in the late 1950s took place in the glare of publicity and public criticism. The Soviets had also been successful in avoiding the interservice rivalry that had erupted in the United States over missiles and satellites. A first step to quashing that rivalry had occurred in 1956 when the U.S. Air Force was assigned long-range missiles, the navy missiles launched at sea, and the army missiles with a range of two hundred miles or less. Still possessing the von Braun team and an aggressive leadership under Major General Medaris, however, the army had not given up. The Army Ballistic Missile Agency put the first U.S. satellite into orbit, and von Braun and Medaris looked to further triumphs in space. Even the creation of the Advanced Research Programs Agency (ARPA) inside the Defense Department in February 1958 and its interim takeover of U.S. space programs didn't discourage them.

The partial separation of America's first satellite program, Vanguard, from the military reflected the views of many in the government and the scientific community that many space activities should be run without military control. Eisenhower himself thought at first that the military could operate the space program as a matter of efficiency, but soon he changed his mind. "Information acquired by purely scientific exploration could and should, I thought, be made available to all the world," he wrote later. "But military research would naturally demand secrecy." The president concluded that the United States should have separate military and civilian space programs. Memories of the battles in the late 1940s over control of America's atomic energy facilities were still fresh, and Eisenhower sought to build a consensus for his space plans by asking James Killian, by then his science advisor, and the President's Science Advisory Committee (PSAC), to design a structure for the space program. Killian, backed up by other scientists, was concerned that a space program controlled by the Defense Department would concentrate on military work and ignore science. Historians David Callahan and Fred I. Greenstein have written that Eisenhower knew that Killian shared his ideas on the space program and that the president's handling of the problem was a good example of his "hidden hand approach to leadership."

Killian and PSAC soon concluded that the new civilian space agency should be established on the foundation of an existing agency, the National Advisory Committee for Aeronautics (NACA), a modest but respected agency that had been set up in 1915 to advance aeronautical research. Funding for NACA had peaked in World War II, but cost-cutting after the war had halved its funding by 1954. Historian Walter McDougall has noted that NACA lacked allies because little of its funding went to aeronautical contractors. Moreover, the agency had deliberately kept its distance from space projects until just before Sputnik, when younger engineers began pressing for more space-related work. By the time PSAC was preparing its report, NACA was lobbying to enlarge its responsibilities in order to become the space agency, and Killian and PSAC were happy to oblige. On April 2, the president sent a message to Congress with a bill establishing the National Aeronautics and Space Agency. Although proposals to keep the space program within ARPA or move it to the Atomic Energy Commission were also floated, Democratic leaders of the House and Senate soon introduced their own bills, which were similar to the administration's proposal. Legislative work began quickly thereafter, and both the House and Senate held hearings and ultimately passed their own versions of the legislation. The House version of the bill upgraded the new agency to an administration, thus putting it higher in the Washington pecking order. When a conference committee made up of members of both bodies hashed out a final bill, NASA officially became an administration rather than an agency, and the Senate got its way on its desire for a National Aeronautics and Space Council. Eisenhower opposed the council as a check on his power, but he gave way when the president was made the council's chair. Both the House and Senate passed the conference bill on July 16, and Eisenhower signed it into law on July 29.

NASA Begins Work

To run NASA, Eisenhower and Killian chose T. Keith Glennan, the president of Case Institute of Technology in Cleveland, and a man who shared the president's concern that government was getting too big. Glennan accepted on the condition that the director of NACA, Hugh Dryden, a widely respected engineer and manager who had himself been considered for the administrator's job, become his deputy. Glennan and Dryden were sworn in on August 18, and on October 1, 1958, NASA opened for business.

On that day, NASA took over the NACA staff of eight thousand, its budget of $100 million, and its facilities, including a small headquarters staff in Washington and three research laboratories: the Langley Aeronautical

24. Dr. Hugh L. Dryden served as NASA deputy administrator from August 19, 1958, to his death on December 2, 1965. Dr. Dryden held the position of director of the National Advisory Committee for Aeronautics (NACA), NASA's predecessor, from 1947 until October 1958. (NASA photo)

Laboratory in tidewater Virginia, the Ames Aeronautical Laboratory south of San Francisco, and the Lewis Flight Propulsion Laboratory in Cleveland. NACA also handed over a flight-test facility for aircraft at Edwards Air Force Base and for small rockets at Wallops Island, Virginia. Langley, Ames, and Lewis were redesignated as NASA research centers, and soon NASA was given control of Project Vanguard's personnel and facilities, which had previously been under the jurisdiction of the U.S. Navy. The Vanguard group became the foundation for NASA's new Robert H. Goddard Space Flight Center, set up in 1960 in Maryland between Washington and Baltimore.

NASA also wanted to take over two institutions that had worked with the army: the Jet Propulsion Laboratory in Pasadena, California, and most of the Army Ballistic Missile Agency. Both groups resisted, both because of loyalty to the army, which still held onto its dreams of space exploration, and doubts about the new space agency's effectiveness in its new mission. In 1958, JPL had gained fame for creating *Explorer 1,* and was looking at building probes that would go into deep space and to the planets. ARPA had already cut back JPL's rocket work, and JPL director William Pickering suggested that the organization should be designated as America's "national space laboratory," so it could spearhead the U.S. space program rather than the former leaders of NACA, who JPL saw as taking a hesitant approach to space. "If this is not done," Pickering said, "then NASA will flounder around for so long that there is a good possibility the entire program will be carried by the military with NASA providing only some research support and perhaps helping with scientific payloads." The army did not want to lose JPL, but its desire to keep von Braun's rocket team at ABMA was stronger. As a result, JPL came under NASA jurisdiction on January 1, 1959, in spite of the JPL leaders' misgivings about the new space agency.

The terms of the original agreement continue into the twenty-first century: JPL remains a part of Caltech, operating under a contract with NASA. Upon joining NASA, JPL quickly completed its remaining rocket work for the army and turned to building spacecraft to explore the moon and the planets. Although its plans at the time foresaw ambitious exploration to the moon, Mars, and Venus beginning in 1959 and 1960, the engineers and scientists at JPL quickly learned some hard lessons about the difficulties of moving out from earth.

The bureaucratic battle over ABMA proved to be more intense than that over JPL. Von Braun let the press know even before NASA officially began operations that he did not wish to be part of the new agency. In meetings with NASA officials, Wilbur Brucker, the secretary of the army, strongly resisted NASA's attempts to take over ABMA, and Medaris spoke publicly about the army's need to keep its missile team. McDougall speculated that von Braun took this attitude out of loyalty to Medaris, and "probably thought that a feeble, civilian agency would never command big money." In the bargaining for JPL, the army was able to delay ABMA's move to NASA, but in September 1959 the Defense Department assigned most space-related missions to the air force. As a result, the army was forced to confront the fact that its rocket team would have little to do. Medaris angrily retired from the army, and the transfer of much of ABMA to NASA was negotiated in October.

As early as 1957, many people in the Defense Department saw the need for a new rocket with a first-stage thrust of 1.5 million pounds to launch military payloads into orbit. Beginning work on the concept, von Braun's ABMA team came up with a rocket whose first stage consisted of a cluster of eight engines built by Rocketdyne for the Jupiter missile that ABMA had built as the follow-up to Redstone. The clustered engines avoided technical problems associated with building a larger engine, and in 1958 ARPA agreed

25. Five pioneers pose with scale models of the missiles they created in the 1950s. From left to right: Dr. Ernst Stuhlinger, a member of the original German rocket team and director of the Research Projects Office, Army Ballistic Missile Agency (ABMA); Maj. Gen. Holger Toftoy, who consolidated U.S. missile and rocketry development; Prof. Hermann Oberth, a rocket pioneer and Dr. von Braun's mentor; Dr. Wernher von Braun, director, Development Operation Division, ABMA; and Dr. Eberhard Rees, also a member of the original German rocket team. (NASA photo)

to continue the rocket studies. By then, ABMA had decided that the tanks should be clustered too. So the rocket, which soon became known as Saturn, had a first-stage structure made up of eight Redstone tanks clustered around a Jupiter tank. By 1959, ARPA questioned the need for such a rocket for military purposes, and funds dried up just as the first models were being built. But NASA foresaw that the Saturn could aid exploration of the moon, and Saturn's survival within NASA gave von Braun and his group more respect for the new space agency's strength.

On July 1, 1960, von Braun's group became the nucleus of the newly established George C. Marshall Space Flight Center, which shared the government land at Huntsville, Alabama, with the army's Redstone Arsenal. The new center was named for General of the Army George C. Marshall, who had also served as secretary of defense and secretary of state. Under von Braun's direction, the new center had primary responsibility for NASA's launch vehicle programs, including the NASA launch operations at Cape Canaveral, Florida. Marshall was able to press on with the development of the Saturn rocket, and a proud von Braun showed Eisenhower the Saturn when the president came to Huntsville to dedicate the new NASA center. The first stage of the Saturn rocket von Braun was building formed the foundation for the Saturn I and IB rockets used in the Apollo and Skylab programs, as well as the gigantic Saturn V rocket later developed for the Apollo moon missions. The Huntsville rocketeers also inherited the gigantic F-1 engine, which the air force had already begun developing when it was shifted to NASA in 1958. The 1.5-million-pounds-of-thrust engine was later clustered in the first stage of the Saturn V.

As for von Braun himself, America's fears over Sputnik, combined with his personal triumph in launching *Explorer 1*, catapulted him to the height of his fame. This new level of notoriety went well beyond what he had attained through his appearances in *Collier's* magazine and on the *Disneyland* television show. Shortly after the success of *Explorer 1*, von Braun appeared on the cover of *Time* magazine, for example, and he was invited to visit the White House to receive congratulations from President Eisenhower. In 1960, von Braun was even the subject of a Hollywood movie, *I Aim for the Stars*. The generally positive treatment of von Braun brought out anti-Nazi critics who picketed theaters in some places. More than one comedian quipped that the film should be called, *I Aim for the Stars, But I Hit London*. Von Braun biographer Dennis Piszkiewicz has written that von Braun himself disliked the film, which contained many subplots created solely for dramatic effect. The film neither reflected his life nor the legend that had built up around it.

In spite of von Braun's successes in winning public acclaim for his work in America, the protesters and the jokes reflected a dark past that always dogged him. Piszkiewicz relates how von Braun once pointed out to a Disney writer the two FBI agents who tailed him constantly, even as he prepared for one of the *Disneyland* shows in the 1950s. Many people whom he worked with in the army and at NASA had fought against Nazi Germany in the war, and von Braun never forgot his obligation to work hard to make a good impression. "He was tarred with the same brush the other Germans were from my experience in World War II," remembered Owen Maynard, a NASA engineer who worked at another NASA center. "In my perspective, all these guys were bad guys, and I had been trained to think that way." When Maynard and other engineers working on the Apollo program were told to meet with von Braun, they did so reluctantly. Upon their arrival at Huntsville, a combination of von Braun's effort to learn their names before the meeting, his charm, and a dash of self-deprecation changed the visiting engineers' minds. "We left liking them more than when we came in," Maynard said. "I was sort of ashamed of myself. That personal encounter was definitely in a favorable direction."

Flying to the Moon

Though NASA was still just being formed, the United States was contending with challenges in space from the Soviet Union as both countries looked to the moon and beyond to score points in the dawning space race. Fresh from the success of *Explorer 1*, America's new and temporary military space agency, ARPA, mandated that the air force and army prepare rockets and spacecraft to beat the Russians to earth's natural satellite. Three lunar probes resembling child's tops were thus built by Space Technology Laboratories of Redondo Beach, California, a subsidiary of TRW. The air force probes were designed to go into lunar orbit and snap a few images of the moon after launch atop a Thor-Able rocket, which consisted of an air force Thor missile with a Vanguard upper stage. On top of building rockets and space probes to go into deep space, engineers now had to develop the means to track space probes at great distances. ARPA authorized the first steps toward America's deep-space tracking network by constructing large antennas at Goldstone, California; near Madrid in Spain; Canberra, Australia; and Johannesburg, South Africa.

Humanity's first attempt to reach the moon left Pad 17-A at Cape Canaveral at 8:18 a.m. on August 17, 1958. But seventy-seven seconds later, a bearing froze in Thor's engine, and the vehicle exploded. When the air force tried again on October 11, all three stages of the rocket worked, and the spacecraft designated after launch as *Pioneer 1* leaped into space. The rocket's aim was slightly off, however, and its performance was slightly below what was needed to get *Pioneer 1* to the moon. As a result, the spacecraft soared 70,700 miles away from earth and then fell back into the atmosphere forty-three hours after launch. Nonetheless, *Pioneer 1* set new altitude and space records, allowed scientists to conduct communications repeater tests, and sent back new information on the earth's newly discovered Van Allen radiation belts. The launch was called a "magnificent failure." The third and final air force lunar launch on November 8, *Pioneer 2*, flew for only forty-one minutes up to 963 miles when the third stage failed to fire.

The next two launches in ARPA's moon program went to ABMA, which used Jupiter missiles fitted with upper stages built by JPL to form a rocket known as Juno II. The cone-shaped spacecraft built at JPL carried Geiger counters to measure radiation, after *Pioneer 1* found higher levels than expected. The hard luck of the air force probes continued when the Jupiter engine burned out nearly four seconds prematurely. Although the remaining two stages worked properly, *Pioneer 3* flew 66,654 miles high and fell back to earth thirty-eight hours after its launch on December 6, having made important measurements of the Van Allen belts along the way. The army had one more chance for the moon, but the probe was upstaged by a launch from America's Cold War adversary.

The Soviet government, having enjoyed the world's reaction to its first two Sputniks, decided in March 1958 to launch spacecraft to the moon. The Soviets ordered the construction of several vehicles, including spherical craft not much bigger than the first Sputnik but now filled with experiments to investigate the gas composition of interplanetary space and to search for radiation, cosmic rays, and magnetic fields. Other spacecraft were designed to photograph the unseen far side of the moon, and the Soviets also considered detonating a nuclear bomb on the moon, as did some people in the United States. This idea was wisely shelved.

The R-7 booster required an additional stage to boost the spacecraft to the higher speeds required for the moon. The first Soviet moon launch attempt took place on September 23, 1958, barely a month after the first American attempt. Both this launch and a second launch a few hours after *Pioneer 1* on October 11 failed during the first two minutes of flight due to problems related to the addition of a third stage to the R-7. The failures continued on the third attempt in December, which could not reach the required velocity. All these launches remained secret for many years.

Finally, on January 2, 1959, the Russians tasted success. A spacecraft variously called *Luna 1*, Cosmic Rocket, or Mechta ("Dream") reached escape velocity and became the first object launched by humans to go into orbit around the Sun. Although *Luna 1* missed hitting its target of the moon, it was a triumph all the same. The spacecraft sent back data until it was well past the moon, and anyone who doubted the Soviet achievement had to account for the sodium cloud that the probe's third stage rocket released at an altitude of seventy-five thousand miles, which was visible from the eastern hemisphere.

ARPA's and the army's last lunar probe, *Pioneer 4*, was launched on March 3, 1959, and this time the effort was crowned with success. The craft flew by the moon and became the first American probe to enter solar orbit. A slight aiming error meant that *Pioneer 4* missed the moon by twice the planned twenty thousand miles. Because its flight was so similar to *Luna 1,* the flight of *Pioneer 4* brought little recognition to the United States. At this point, ARPA left the space field but continued work in other areas, including computer research, where a networking effort that began in 1969 led to the agency's most famous invention, the Internet.

The Soviets weren't done with the moon. On September 12, *Luna 2* was launched from Tyura Tam, and two days later, it became the first human object to strike another celestial body when it crashed on the moon near the Sea of Serenity. This achievement took place on the eve of Khrushchev's first visit to the United States, and the gleeful Soviet premier took delight in presenting Eisenhower with a replica of the pennant bearing Soviet emblems that *Luna 2* carried to the lunar surface.

Three weeks later, on the second anniversary of Sputnik's launch, Radio Moscow broke into its telecast with this dramatic announcement: "Attention, attention, dear Comrades! Listen now to the signals from the cosmos, from the third cosmic rocket launched today." The radio then broadcast the sound of the telemetry coming from *Luna 3*. This new probe, like its predecessors sometimes known by its nickname, Lunik, was new in that it was designed not to hit the moon but to get photos of its far side, never before seen by human eyes. Its booster accurately injected *Luna 3* into a highly eccentric earth orbit that took it near the moon. When the probe passed around the far side, it took twenty-nine photos and developed them. Scans of the film were broadcast back to earth when *Luna 3* passed closer to home. Though the vehicle encountered problems with overheating and transmitting to earth, seventeen photos contained usable data. The photos, which showed 70 percent of the moon's far side, were of poor quality, but they unmistakably demonstrated that the moon's far side had far fewer dark "seas" than its earth-facing side. Two small "seas" were found among some large craters, notably a spectacular crater filled with dark material that was named after Konstantin Tsiolkovsky. The first generation of Soviet lunar craft concluded in 1960 with two failed launch attempts of improved versions of *Luna 3*. The second launch attempt failed shortly after liftoff, shattering windows and sending people scattering from the Tyura Tam launch pad.

By the time these Soviet triumphs had impressed the world, NASA was ready to try new Pioneer class spacecraft to fly into lunar orbit. These spherical spacecraft, which were powered by solar cells on four paddles, were launched on Atlas missiles with Able upper stages. Three attempts in November 1959, September 1960, and December 1960 all failed during the launch phase. These failures marked the end of the first phase of American lunar exploration. By this time, JPL engineers were working on an ambitious new spacecraft to land on the moon called Ranger. The Soviets meanwhile turned to designing a new generation of spacecraft to soft-land on the moon and orbit it.

Both space superpowers were also looking beyond the moon. Two of the final Pioneer-class spacecraft built by the United States were originally designed to fly to Venus before they were diverted to missions to lunar orbit. Another Pioneer was launched atop a Thor-Able booster on March 11, 1960, for a flight in solar orbit. Scientists wanted to use *Pioneer 5* to learn about the magnetic and radiation environment beyond the earth-moon system, and the craft functioned for 109 days, providing data from a record distance of 22.5 million miles. JPL engineers then began work on a new class of spacecraft called Mariner to fly to Venus and Mars in the 1960s.

The Soviets were also aiming for Venus and Mars. They set to work building larger antennas on Soviet territory to track these spacecraft, and designed vehicles with solar panels that would last more than the few days required to get to the moon. Because of the planets' motions relative to earth, launch opportunities for both planets are limited. In the case of Mars, these opportunities arise only for a short period every two years. When such a launch window arrived in October 1960, the Soviets sent two rockets toward the Red Planet, but neither made it into parking orbit around the earth. Khrushchev was then in New York to speak at the United Nations, and a sailor who defected from the ship that brought the Soviet leader across the Atlantic told the press that the ship carried a model of the Mars spacecraft.

In February 1961, an opportunity to fly to Venus opened, and two Venera spacecraft were launched. Only the second craft, which was called *Venera 1*, was successfully injected into a path to Venus, on February 12. But after a communications session five days into flight the spacecraft was never heard from again, and it flew silently by the veiled planet in May. Like the Americans, the Soviets planned new flights to both planets later in the 1960s.

Applications Satellites

While the moon shots dominated the headlines in 1958 and 1959, a more immediately important part of humanity's move into space was gaining momentum, particularly in the U.S. space program. These were the satellites that were designed to serve earthly purposes, such as improving communications links, predicting the weather, and assisting navigators. Arthur C. Clarke had already forecast the potential of geosynchronous communications satellites in 1945. As the space age opened, he had written: "In a few years every large nation will be able to establish (or rent) its own space-borne radio and TV transmitters, able to broadcast really high-quality programs to the entire planet." This advance, he wrote, would "mean the end of all distance barriers to sound and vision alike."

In the 1950s, communications scientists were bouncing radio signals off the moon, and one of the first prominent programs to emerge from the newly established NASA was a test of communications technology called Project Echo. Based on an idea by Langley engineer William J. O'Sullivan, *Echo 1* was a 100-foot-diameter, aluminum-coated Mylar balloon called a "satelloon" that NASA put into orbit atop a Thor-Delta rocket on August 12, 1960. NASA engineers at Langley and Goddard Research Centers built and tested the balloon, which became famous because of its great visibility from the earth during the eight years it was in orbit. Researchers bounced radar and communications signals off the *Echo 1*, including a message from President Eisenhower, but they found that passive reflectors were of limited value. Shortly after Echo was launched, the U.S. military began testing active repeater communications satellites in low orbits. These satellites contained equipment that amplified communications signals before they left the satellite on their way to target antennas on earth. Echo was important because it had opened the minds of many people to the idea of communications satellites. These would grow into the active repeaters in geosynchronous orbits, becoming within a few years a crucial part of the world's communications infrastructure. Scientists also learned about the composition of the upper atmosphere by following its effects on Echo. *Echo 1* also demonstrated how satellites could be used to determine the distances between various points on earth through triangulation, an insight that led to more accurate mapping of the earth. These concepts were further tested in a second Echo satellite and other satelloons.

Another important initiative from NASA in 1960 was the first weather satellite, TIROS 1 (for Television Infrared Observation Satellite), which was launched on April 1 by the final Thor-Able rocket and sent back its first photo that day. TIROS quickly proved the value of satellite observations in weather forecasting, and continuous service by these satellites began two years later. That same month, the U.S. military also launched the first Transit navigation satellite, which evolved into today's Global Positioning System (GPS) satellites. Within a few years, satellite observations became an indispensable part of weather forecasting, and by the turn of the century GPS systems were becoming an important part of everyday life.

International Cooperation

The early years of the space age are remembered for the U.S. competition with the Soviet Union, but the first glimmers of cooperation that would become increasingly common and important in space exploration late in the twentieth century also date to this time. The law creating NASA, the National Aeronautics and Space Act of 1958, permits NASA "to engage in a program of international cooperation in work done pursuant to this Act, and in the peaceful application of the results thereof." NASA soon began discussions with close allies such as Canada and the United Kingdom that led in 1959 to joint programs whose culmination in the 1960s was the launch of Canadian Alouette and British Ariel satellites atop NASA rockets. These initiatives led to agreements with other countries in the years to come.

The International Geophysical Year had seen a great deal of cooperation between scientists from around the world. One of its legacies was the Committee on Space Research (COSPAR), which still operates under the auspices of the International Council of Scientific Unions as a forum for space scientists, advisory body on space to the United Nations, and promoter of space research. The committee was set up as a means of continuing work begun in the IGY, and although Cold War differences between the United States and Soviet Union gave COSPAR a difficult birth, these were settled in 1960.

These differences also meant a rocky start for the United Nations Ad Hoc Committee on the Peaceful Uses of Outer Space (COPUOS), which was created by the UN General Assembly in the fall of 1958. The committee's original mandate reflected the ideas of Western countries because a Soviet proposal for the committee was defeated. Nevertheless, COPUOS became the forum where space law was drawn up in the coming years, notably through the Outer Space Treaty of 1967.

The high level of international cooperation in space at the beginning of the space age reflected the fact that scientists and rocket engineers from many nations, not merely the United States and Soviet Union, were at work. In the 1950s, Europe was still recovering from the devastation of World War II but was already looking to new advances in science. The United Kingdom had a long history of scientific work in astronomy and upper-atmosphere research, and British scientists took an active part in the IGY. In 1955, the British government began developing its own IRBM, known as the Blue Streak, in cooperation with the United States. Though Germany avoided rocket development after the war as it rebuilt itself under restrictions on military research, some veterans of Peenemünde who remained in Germany encouraged space research. About forty Peenemünde experts who moved to France after the war helped establish rocket research there, for example. French space efforts operated on a modest scale until 1958, when the launch of Sputnik and the return to power of Charles de Gaulle led to an acceleration of rocket research in support of de Gaulle's ambitions to make France a nuclear power. This policy led to the creation of a French family of sounding rockets and launch vehicles.

European scientists were already working together across national borders through the IGY and through European institutions such as the CERN nuclear particle accelerator, which began operations in Geneva in 1959. All this took place alongside wider political moves to create today's European Union. Scientists such as Italian physicist Edoardo Amaldi; French physicist Pierre Auger; and Harrie Massie, the head of Britain's space research committee began a series of discussions on forming a new European space organization. In 1960, the British decided to cancel the use of Blue Streak as a weapon and instead proposed that it form the basis of a European space launch vehicle. These discussions led to the creation of the European Space Research Organization and the European Launcher Development Organization in the 1960s and the European Space Agency in the 1970s.

China's rocket and space program began with Tsien Hsue-shen's return to China in 1955, after his distinguished rocket career at Caltech and MIT was cut short by wild anticommunist accusations against him in the early 1950s. Though Tsien was not allowed to bring back his papers, he replicated this work from memory. China in 1955 was still emerging from decades of war and political turbulence that climaxed with the communist takeover in 1949. China's communist leader, Mao Zedong, was anxious to develop science and technology in China, so in 1956 the government established the Fifth Research Academy of the Ministry of National Defense to run China's rocket and space programs. Along with other scientists, some of whom were also trained in the United States, Tsien began teaching the science of rocketry to build up China's expertise.

More help was needed, however, so the Chinese government began talks with the Soviet government that year to share rocket technology. The two communist countries already cooperated in several areas including nuclear weapons, and a deal covering rockets was signed in 1957. The Soviets sold the Chinese their R-1 rocket, the Soviet copy of the German V-2, and then their R-2. Soviet experts also came to China, while Chinese graduate students went to study in Moscow. Impressed by Sputnik, Mao approved a Chinese satellite program in 1958, but the program was shelved a few months later when the Fifth Academy grasped the difficulties that lay before it. China reeled from the failure of the Great Leap Forward, Mao's scheme to accelerate the Chinese economy, which instead led to economic devastation and widespread famine.

In spite of this setback, the Chinese built a sounding rocket and established a permanent launch site at a desolate spot in the Gobi desert at

Jiuquan. Another setback arrived in August 1960 when the once "unshakable" alliance between the Soviets and the Chinese broke down, apparently over differences concerning nuclear weapons. The Soviet advisors abruptly canceled joint projects and went home. But by then, the Chinese had succeeded in building their own version of the Soviet R-2, a rocket they called Dong Feng 1 or East Wind. On November 5, the first Dong Feng 1 rocket flew successfully, and two further successful flights confirmed that China had its own workable rocket. This success led to new and larger rockets over the succeeding decades, but the program continued to face political challenges as China underwent further political turmoil in the 1960s and 1970s. China emerged as a major space power in the 1980s, and early in the twenty-first century the first Chinese taikonaut soared into the skies from Jiuquan, with Tsien still alive to see his dream realized.

Two other future space powers—Japan and India—were still on the sidelines when Sputnik flew. India's top space pioneer, physicist Vikram Sarabhai, conducted cosmic ray research in 1950s. Impressed by the work done during the IGY, he pressed to organize the building of sounding rockets in the 1960s, which eventually led to larger rockets and satellites. In Japan, Prof. Hideo Itokawa's launches of tiny "pencil" rockets in 1955 and Japanese participation in the IGY led to satellite launchers and spacecraft in the 1970s.

Canada had a long history of upper-atmosphere research, bolstered in the 1950s by American sounding rocket launches from Fort Churchill in northern Manitoba. Canada developed its own sounding rocket, the Black Brant. In 1959, under Dr. John Chapman, Canada started work on its own satellite, *Alouette 1*, which furthered upper-atmospheric research after NASA launched it in 1962.

Most of these nations' interest in entering space reflected the hope of achieving benefits on earth. Chapman, for example, expressed the wish that "the fabric of Canadian society will be held together by strands in space just as strongly as railway and telegraphy held together the scattered [Canadian] provinces in the last century." Sarabhai believed that India and other developing countries needed to harness advanced science, including space science, to fight poverty. "There are some who question the relevance of space activities in a developing nation," he said. "To us, there is no ambiguity of purpose. We are convinced that if we are to play a meaningful role nationally, and in the community of nations, we must be second to none in the application of advanced technologies to the real problems of man and society."

Nuclear Rockets

Aside from space, the science of the 1950s was also dominated by research into things nuclear, including weapons and power sources. The development of rockets and satellites made it almost inevitable that both space and nuclear research would come together at some point in the decade. The U.S. Atomic Energy Commission began Project Rover in 1955 to develop nuclear rocket engines that would be powered by hydrogen gas superheated by passage through nuclear reactor cores. Scientists at the Los Alamos Laboratory began design work for nuclear reactors and engines for Rover, including KIWI, Pewee, and later NERVA or Nuclear Engine for Rocket Vehicle Applications. A number of these engines were tested at the commission's test site in Nevada. The funding for this effort dried up in the late 1960s and early 1970s.

Another proposal known as Project Orion began in 1957 from a concept of nuclear weapons designer Theodore B. Taylor, then at General Atomics, a firm developing nuclear reactors in San Diego. Taylor recruited a group of scientists, notably physicist Freeman Dyson from the Institute for Advanced Research in Princeton, New Jersey, to work on Orion. Taylor and his Project Orion team based their work on an idea by physicist Stanislaw Ulam to eject nuclear bombs from behind the spacecraft and then use the explosions to strike a pusher plate that would propel the spacecraft. Starting with modest funding from ARPA in 1958, Project Orion began testing the idea of pusher plates and designing huge spacecraft that could fly to distant points in the solar system. When ARPA got out of the space business after the formation of NASA, Project Orion continued with funding from the U.S. Air Force. Despite von Braun's interest in Orion, its leaders could not persuade NASA to support the program, and it expired in 1964 from lack of funds. Since then, there has been discussion of but little serious action toward the development of nuclear power engines that could exceed the performance of conventional chemical rocket engines. Aside from small experiments with ion or other exotic propulsion sources, chemical-based rockets remain the sole propulsion source for space vehicles at the beginning of the twenty-first century. Still, the dream of nuclear propulsion remains alive.

As Dyson wrote: "The men who began the project in 1958 aimed to create a propulsion system commensurate with the real size of the task of exploring the solar system, at a cost which would be politically acceptable, and they believe they have demonstrated the way to do it." Unfortunately, those who funded the programs ultimately chose to disagree.

The first three years of the space age had seen great progress in many countries. As the third anniversary of Sputnik was marked in 1960, Eisenhower's Republican vice president, Richard Nixon, was locked in a tight battle in the presidential election with Democratic senator John F. Kennedy. "The first man-made satellite to orbit the earth was named Sputnik," Kennedy warned. "The first living creature in space was Laika. The first rocket to the moon carried a red flag. The first photograph of the far side of the moon was made with a Soviet camera. If a man orbits the earth this year, his name will be Ivan." Nixon responded with the fact that the United States had successfully launched twenty-six satellites into earth orbit and two spacecraft into orbit around the Sun, versus only six satellites and two successful space probes by the Soviets. Although the Soviets had won the headlines, the United States was far ahead in building the necessary infrastructure to make space useful for people.

Moreover, in spite of Nikita Khrushchev's famous boast that the Soviet Union was making ICBMs "like sausages," the Soviet R-7 was not well suited for use as a weapon. The Americans were already deploying Atlas and more advanced Titan ICBMs, and solid-fueled Minuteman missiles were on the way. On October 24, 1960, as the U.S. presidential election was entering its final days, a test of an R-16 ICBM at Tyura Tam turned into a catastrophe shortly before the scheduled launch. The missile, comparable to an American Titan II, exploded on the pad while many technicians and high officials were still nearby. The accident took the life of 126 people, including Marshal Mitrofan Nedelin, the head of the Soviet Strategic Missile Forces, and many top designers. Fortunately for the Soviets, their top ICBM designer, Mikhail Yangel, had chosen that moment to take shelter in a bunker and have a cigarette, a decision that saved his life. The facts about this terrible setback for the Soviet missile and space program were kept secret for nearly thirty years, and ultimately the R-16 and other rockets designed by Yangel came to form the core of the Soviet ICBM force.

Yangel's growing importance to the Soviet military stood in contrast to Korolev, who barely concealed his preference for working on space exploration over developing missiles for military use. "Korolev works for TASS [the Soviet news agency], Yangel for us," was a popular saying in the military at the time. The R-16 rocket disaster caused only a pause in the Soviet missile and space efforts. Soon Yangel was back perfecting the R-16, and Korolev turned to his next goal: making Kennedy's pessimistic prediction about the nationality of the first human in space a fact.

11. Man in Space Soonest

We are rapidly approaching the time when the state of technology will make it possible for man to go out into space. It is sure that, as soon as this possibility exists, man will be compelled to make use of it, by the same motives that have compelled him to travel to the poles and to climb the highest mountains of the Earth. There are also dimly perceived military and scientific missions in space which may prove to be very important.

Report to the President-elect (John F. Kennedy) of the Ad Hoc Committee on Space ('The Wiesner Report'), January 1961

The launch of the dog Laika aboard *Sputnik 2* in November 1957 was an unmistakable signal that the day was not far off when other living beings—this time human—would fly into space. The fact that Laika had survived her first hours in space answered the first questions about whether humans could survive the rigors of a spaceflight. Any living being going into orbit had to contend with acceleration forces during the launch and then weightlessness in orbit. How weightlessness would affect bodily functions was a big mystery in 1957. There was also the matter of developing systems that could support life inside sealed spacecraft. And the big issue that remained unanswered by *Sputnik 2* was whether living beings could be brought back to Earth safely.

Many people were hard at work on these questions even as the two sides in the Cold War struggled to get their first satellites into space. In the United States, the air force had begun immediately after the war to explore the physiological issues posed by high-speed and high-altitude flight at the Aeromedical Laboratory at Wright-Patterson Air Force Base in Dayton, Ohio. At the School of Aviation Medicine at Randolph Air Force Base in San Antonio, Texas, the air force brought together a number of German physicians, headed by Hubertus Strughold, who had done research for the Luftwaffe during World War II. In 1949, the school set up the first department of space medicine under Strughold, and doctors there and at the Lovelace Foundation for Medical Research in Albuquerque, New Mexico, conducted research on some of the human problems related to going into space. Strughold was later honored as the father of space medicine until his role in inhumane experiments on concentration camp inmates during World War II came to light.

The postwar medical work in the United States centered around launches of animals atop V-2 and Aerobee rockets, the rocket sled and balloon efforts run by Col. John P. Stapp, and the flights in the X series of experimental aircraft. Brother physicians Fritz and Heinz Haber of the Air Force School of Aviation Medicine were looking for a means of simulating weightlessness in space and decided that the best way was to fly aircraft along vertical parabolic trajectories. In 1951, air force test pilot Chuck Yeager and NACA test pilot Scott Crossfield took a break from their informal competition for speed and altitude records to test the Habers' theory in fighter aircraft, and both pilots experienced brief periods of weightlessness. These initial tests led to more ambitious flights in different aircraft, in which test subjects determined how weightless conditions would affect different bodily functions.

Physicians had begun studying the effects of high acceleration on humans during World War II by observing pilots subjected to high "g" forces in fighter aircraft, and later test pilots in the new jet and rocket aircraft. Stapp and his associates tested the capacities of humans and animals to survive high acceleration and deceleration. By the mid-1950s, however, it was clear that another research tool was needed to test the challenges faced by jet pilots in their aircraft and ejection scenarios, not to mention the human-factors questions for pilots flying beyond the atmosphere. The needed research tool was the centrifuge, which had first been used by the German Luftwaffe during the war. At Wright-Patterson in 1935 the U.S. Air Force built a centrifuge, which consisted of a long arm with a gondola at the end that could whirl its passengers at high angular velocities, and the U.S. Navy built one at its Aviation Medical Acceleration Laboratory in Johnsville, Pennsylvania. For the rest of the decade, researchers at both facilities determined how the human body could best accept acceleration and deceleration forces during launch and reentry. Some experiments involved immersing test subjects in water tanks while they sustained high acceleration loads. Researchers in America, Canada, and Germany tested other devices such as hammocks and special couches, but in the end they found that being semisupine was the best position in which to deal with the forces of acceleration and deceleration.

Another human-factors issue was designing sealed cabins and environmental control systems that supplied air and disposed of exhaled carbon dioxide. The special chambers built at the aviation medicine facilities together with the experience of high-altitude balloonists and the sealed cabins used in undersea exploration helped engineers and doctors cope with the problems of sealed environments. The possibility of puncture by micrometeoroids and exposure to radiation in space also concerned those preparing for the first spaceflights.

As part of efforts in 1958 to prepare for space travel, Americans launched animals atop missiles for the first time in six years. Three small white mice were launched on separate flights in the nose cones of Thor-Able rockets. That same year the army's surgeon general asked the Army Ballistic Missile Agency to make room available in missile nose cones for spaceflight experiments involving animals. On the first of these flights, Friday, December 13, 1958, a South American squirrel monkey named Gordo was launched aboard a Jupiter missile. Gordo made an apparently successful quarter-hour flight through space, with an estimated nine minutes of weightlessness. He is thought to have survived the reentry flight, but plans for his Atlantic Ocean recovery were abandoned when the nose cone's signal beacon failed to transmit. On May 28, 1959, another Jupiter launched a rhesus monkey named Able and a squirrel monkey named Baker, along with other biological specimens. The Jupiter carried the nose cone nearly 225 miles into space, reaching a speed of 11,250 miles per hour and traveling 1,560 miles down range. Following the ballistic flight, the nose cone separated as planned and parachuted down to the ocean, and the monkeys were safely recovered. Able died shortly after recovery while electrodes were being removed from her body, but Baker, also known as Miss Baker, lived on for twenty-five years as a beloved resident of the Alabama Space and Rocket Center in Huntsville.

More Balloon Flights

The launch of Sputnik breathed new life into the air force's Project Manhigh balloon program, which after David Simons's August 1957 flight in *Manhigh II* had been winding down with little support. Simons, who had also made major contributions to space medicine with his work in the Project Blossom animal flights and the early airborne simulations of weightlessness, was ordered to proceed with a third Manhigh flight. It took place on October 8, 1958, with air force Lt. D. Clifton McClure on board. McClure was forced to cut his flight short when he was just shy of 100,000 feet after his cooling system failed. He fought off life-threatening heat inside his tiny cabin to complete the flight, and the Manhigh program, successfully. The navy, for its part, was continuing with its two-man Strato-Lab balloon flights. The fifth and last Strato-Lab flight reached 113,000 feet on May 4, 1961.

In early 1958, Stapp, who had started Manhigh, and Joseph Kittinger, who had flown *Manhigh I*, were both at Wright-Patterson working on problems related to bailouts at high altitudes. To deploy their parachutes, pilots bailing out needed new equipment that would enable them to escape the flat spins they tended to fall into when ejecting at high altitudes. In Project High Dive, the air force had already started using parachuted dummies to find a means to stabilize ejected pilots. Stapp and Kittinger decided to step in where High Dive had left off with Project Excelsior. High Dive's results suggested that a multistaged set of small stabilization chutes deployed at high altitudes were the solution. Kittinger and two colleagues successfully tested the complicated parachute system design in jumps from aircraft at 30,000 feet in October 1959.

Next, Kittinger was cleared to test the parachute system in a jump from a balloon at 60,000 feet. On November 16, he donned a partial pressure suit similar to those used by jet fighter pilots but supplemented with extra layers for protection from the cold of high altitudes. He seated himself in an open gondola, and his balloon, *Excelsior I*, took off from a launch site at Truth or Consequences, New Mexico. On the way up, however, the sun blinded him, his helmet visor fogged up, and his helmet began to ride up as if to come off—posing fatal consequences. Kittinger was already a mile above his planned jump altitude when he discovered he couldn't free himself from his seat. Finally, he was able to loosen his instrument pack and stand up, but he still had to fight with a stuck lanyard. At 76,000 feet, Kittinger jumped. He was surprised to find that the sensation was like hanging still in space. His suit didn't ripple. He was so high there was almost no air to give him the sensation of falling, which in fact he was, at four hundred miles per hour. Then the thickening atmosphere yanked him into a spin, and he felt himself blacking out. Kittinger's next sensation was regaining consciousness on the desert floor. His struggle with the lanyard had set a parachute timer off early, releasing a pilot chute prematurely, but his reserve chute had been able to deploy, saving his life. Kittinger was lucky to have survived. In the days that followed, he and the Excelsior team worked to solve the equipment problems that had nearly killed him. On December 11, Kittinger climbed aboard *Excelsior II,* rose to the planned altitude of 74,700 feet, got to his feet, and jumped. This time, everything worked perfectly.

Project Excelsior's goal was to jump from 100,000 feet, however, so on August 16, 1960 Kittinger boarded *Excelsior III* at the launch site in Tularosa, New Mexico. His balloon took off at 5:29 that morning, and as he rose Kittinger noticed that the clouds were thicker than had been forecast. His radio malfunctioned, though he had a backup system. Then halfway up to the jump altitude, Kittinger found that the glove on his right hand wasn't pressurized. A tube feeding oxygen to the glove had failed. As Craig Ryan recounts in his study of the balloon flights, Kittinger faced a choice of coming home immediately and possibly seeing Project Excelsior's goal thwarted, or jumping at altitude and permanently damaging his hand. He chose to ride the balloon to altitude, and at 102,800 feet—just under twenty miles high and above 99 per cent of the atmosphere—Kittinger had time to reflect on the vista above and below him. "As you sit here, you realize that Man will never conquer space," he later wrote. "He will learn to live with it, but he will never conquer it." As the final countdown to the jump began, Kittinger informed ground control of the problem with his right hand. After standing up, he moved to the edge of his gondola and looked down at the clouds below and a sign on the doorstep: "Highest Step in the World." After turning on the movie cameras and voice and data recorders that would follow him down, he took a deep breath at 7:10 a.m. and spoke into his recorder: "Lord, take care of me now." With that, he fell forward from his gondola. As he fell and felt himself hanging in space, he spun around and saw *Excelsior III* receding at a high rate of speed. Sixteen seconds into his jump, Kittinger felt the first stabilization chute deploy, which gently pointed his feet down. He soon was falling near the speed of sound. Four and a half minutes into his jump, he had reached 18,000 feet and the top of the clouds when his main parachute deployed. As he descended under his main chute, he repeatedly thanked God. Because his swollen right hand was immobilized, he couldn't release the instrument pack containing the recorders. He landed hard but safe. Hours later, his hand returned to normal.

The air force provided only minimal funds for the balloon programs, and so it canceled a fourth Excelsior flight since Kittinger had proved the effectiveness of the stabilization parachute system for high-altitude jumps. An idea that Kittinger ride a Redstone rocket and then bail out at an even higher altitude never got beyond the talking stage. Kittinger, who made one more trip to the stratosphere aboard the two-man Stargazer astronomical research balloon late in 1962, is best remembered for setting an altitude record for a parachute jump that still stands today, early in the twenty-first century. Although a handful of balloons with humans on board flew to high altitudes in the 1960s, the completion of Manhigh and Excelsior marked the effective end of an historic period in which small teams of people, led by people like Stapp and Simons, sent balloons to altitudes on the edge of space. Their exploits gave way to larger teams who built aircraft and spacecraft that carried people much higher.

The X-15 and the X-20

Though the Manhigh and Excelsior balloon flights were important, they were tiny efforts compared to the programs to build an aircraft that that could set new speed records, penetrate space, and follow up on the early X-1 and X-2 rocket planes. For years, studies and proposals by NACA and aircraft companies had accumulated, including a 1952 concept by Bell Aircraft consultant Walter Dornberger, the former German general who ran the V-2 development effort. Finally, in 1954 the U.S. Air Force, the navy, and NACA agreed to build a hypersonic rocket plane capable of reaching seven times the speed of sound. Nearly a year later, North American Aviation of Los Angeles beat Bell to win the contract for the plane, which became known as the X-15.

The design and construction process for the X-15 involved many technological advances, not the least of which was developing the exotic materials that would protect the aircraft against the heating effects of high-speed flight. Then there were its rocket engines. The early X-15 flights had to use two XLR-11 engines, the same engine that had powered the X-1, until the more powerful XLR-99 engine was ready. North American designed and built three X-15s, and more than two million hours of engineering effort and four thousand hours of wind-tunnel work were required to make them ready to fly. In his book on Harrison "Stormy" Storms, who headed North American's design team, writer Mike Gray described the X-15 as a "brutal black dart that could easily have been mistaken for a missile except for the slit windows near the nose. Everything about it reeked of speed, from the mirror surface of its black steel fuselage to the tiny wings that were not wings at all but slim steel razors." The X-15 was designed to fly so high that it was equipped with thrusters so it could change its orientation above the atmosphere.

The plane that was sometimes labeled America's first spaceship was rolled out of its manufacturing plant on October 15, 1958. After being moved to Edwards Air Force Base and undergoing ground testing, the first X-15 took to the air under the wing of a B-52 on March 10, 1959. After several captive flights with its launch aircraft, the X-15 was released for its first glide flight on June 10, and its first powered flight with the XLR-11s took place on September 17. At the controls for all these flights was legendary test pilot Scott Crossfield, who had left NACA to join the North American team developing the new aircraft. On his fourth powered flight, Crossfield's engine exploded, but he managed to bring the X-15 home despite a heavy landing that caused the plane's structure to buckle. On June 8, 1960, Crossfield was sitting in the X-15's cockpit during a static test when the XLR-99 engine exploded, seriously damaging much of the aircraft, but not the cockpit or Crossfield.

Even with the less powerful engine, the X-15 had broken Iven Kincheloe's altitude record of 126,200 feet in the X-2 and Milburn Apt's speed record of 2,094 miles per hour, also in the X-2. Crossfield piloted the first flights of the X-15 with its throttleable XLR-99 engine in late 1960 and then handed it over to the air force, navy, and NASA pilots who flew the X-15 until the program ended in 1968. Pete Knight flew the X-15 to a speed of 4,520 miles per hour or Mach 6.7 in 1967, a record for a piloted aircraft in powered flight that still stands. Joe Walker flew the X-15 to an altitude of 354,200 feet or 67.08 miles in 1963, a record for aircraft that stood until SpaceShipOne

26. On September 7, 1956, Capt. Iven C. Kincheloe (left) became the first pilot to climb above 100,000 feet as he rocketed to a peak altitude of 126,200 feet. Just twenty days later, making his first flight in the X-2, Capt. Milburn G. "Mel" Apt (in cockpit) became the first person to exceed Mach 3. He was killed moments later when the aircraft broke up as it tumbled violently out of control. Kincheloe was killed in the crash of an F-104 on July 26, 1958. (U.S. Air Force photo)

broke it in 2004. Only one pilot, Mike Adams, died in an X-15, and several X-15 pilots won astronaut wings for flying higher than 50 miles (the edge of space is today generally acknowledged to be 100 kilometers or 62 miles up). Two X-15 pilots, Neil Armstrong of NACA and Joseph Engle of the air force, later joined the NASA astronaut corps. "In 199 flights over nearly a decade, it would become the most successful research airplane in history," Armstrong later wrote, calling the X-15, "a large ring of keys for unlocking the mysteries of future flight."

The X-15 didn't represent the full extent of air force ambitions in the late 1950s, however. Indeed, the air force invented the term *aerospace* to rationalize its ambitions to control "the total expanse beyond Earth's surface." Studies existed on paper for an upgraded X-15 that could fly higher and faster, but even before the X-15 took to the skies, the air force was putting contractors to work on another craft, the X-20 Dyna-Soar, short for "Dynamic Soaring." Dyna-Soar originated with the Sänger-Bredt antipodal bomber of the 1930s. Walter Dornberger and Krafft Ehricke of Bell Aircraft had proposed

27. NASA pilot Neil Armstrong is seen here next to the X-15 rocket aircraft number 1 after a research flight in California. Armstrong flew the X-15 and other aircraft for the National Advisory Committee on Aeronautics (NACA) and NASA before joining the NASA astronaut corps in 1962. (NASA photo)

a rocket-launched winged bomber in the early 1950s known as BoMi (for Bomber Missile), and when it was revised in 1955 as a reconnaissance vehicle with weapons capability, it was code-named Brass Bell. Other aircraft companies were given study contracts for a similar craft known as RoBo (for Rocket Bomber) and for a research program called HYWARDS (for Hypersonic Weapons Research and Development Supporting System). After months of reviews, the three concepts were fused into one program a few days after Sputnik was launched. In December, the air force issued specifications for Dyna-Soar that envisioned a glider with swept wings and a single pilot that would be tested in a series of drop tests followed by orbital flights atop a Titan rocket. In late 1959, the air force named Boeing as the main contractor. Throughout its history, the Dyna-Soar program suffered from unclear and sometimes conflicting objectives as well as funding problems, and it was ultimately canceled in late 1963 before it could fly. Eisenhower and his successors didn't see a military rationale for putting humans into space, and so the primary human spaceflight programs went to NASA in spite of periodic air force efforts to set up a military human space program. Although Dyna-Soar fell victim to this policy, the work done there included early tests of advanced concepts that would be applied to the space shuttle.

Blunt Bodies

Among the developments that helped undermine winged vehicles such as Dyna-Soar were the creation of ICBMs that could deliver weapons rapidly and precisely to their targets and reconnaissance satellites that could conduct their missions without humans on board. There was another factor, aerodynamics, that unexpectedly conspired against winged spacecraft. As the sound barrier was being broken for the first time, researchers at NACA's Ames Laboratory in California built a wind tunnel that could simulate conditions at speeds well beyond the speed of sound. And like the primitive wind tunnel the Wright Brothers built as they designed their historic aircraft, the supersonic wind tunnel at Ames taught some important and unanticipated lessons. The problem that confronted Ames researchers led by H. Julian Allen in the early 1950s was warheads for ICBMs. Conventional thinking and computer studies suggested that needle-nosed cones were the best design for warheads intended to fall through the atmosphere to their targets. But Ames's wind tunnel research showed that such warheads absorbed so much heat they disintegrated. Though some thought that this made ICBMs useless for delivering weapons, Allen began doing calculations that went beyond what had been done before. "Half the heat generated by friction was going into the missiles," Allen said. "I reasoned we had to deflect the heat into the air and let it dissipate. Therefore streamlined shapes were the worst possible; they had to be blunt." In 1952 Allen and his team found that when blunt bodies struck the atmosphere at high speeds, a shock wave formed in front of them that carried off much of the heat. A great deal of heat still got through the shock wave to the body, so researchers began looking at metals that could absorb the heat, known as heat sinks. They also examined ablative materials that would burn away and carry away the heat. Research on experimental nose cones launched on Jupiter-C and other rockets as well as further wind-tunnel work by NACA researchers at the Ames and Langley laboratories pointed toward ablative materials as offering the best protection for warheads—and spacecraft—descending through Earth's atmosphere.

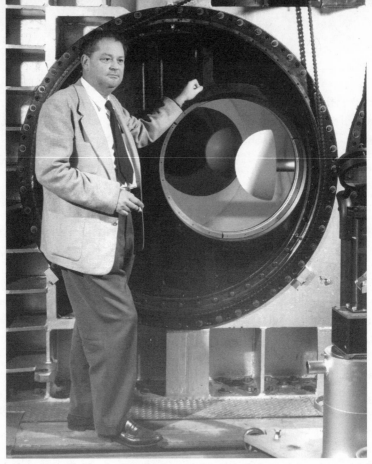

28. H. Julian Allen is best known for his "Blunt Body Theory" of aerodynamics, a design technique for alleviating the severe reentry heating problem that was then delaying the development of ballistic missiles. Subsequently, applied research led to applications of the "blunt" shape to ballistic missiles and spacecraft designed to reenter the earth's atmosphere. (NASA photo)

The shock of the first two Sputniks in 1957 caused anyone in a position to launch people into space to pull together their plans. In the early months of 1958, all three branches of the U.S. military advanced plans for space-flight. The air force was first and most aggressive on this front, starting the year with a five-year plan to put men into space and send them all the way to the moon. The air force also looked at a variety of proposals from NACA as well as from aircraft and missile companies that included the uprated X-15, an accelerated Dyna-Soar, and ballistic capsules based on the blunt-body concept. By spring, the air force's plans had evolved into a series of flights beginning with Man In Space Soonest (MISS), a ballistic capsule that would carry instruments, primates, and finally men into orbit, and ultimately leading to Manned Lunar Landing and Return (MLLR). The Army Ballistic Missile Agency also weighed in with a plan, first called Man Very High and then Project Adam. As the former name suggests, the plan utilized a small capsule similar to the sealed gondola used in the Manhigh balloon flights combined with a Redstone that would launch the capsule and its passenger on a suborbital flight. The navy Bureau of Aeronautics proposed a more sophisticated plan known as Manned Earth Reconnaissance in which a controllable spacecraft would be launched into orbit. All these plans went to ARPA, which controlled military space programs in 1958. But that August, with the NASA bill passed and lacking an immediate and clear military rationale for sending humans into space, President Eisenhower decided to assign the U.S. man-in-space program to NASA, although the air force was allowed to continue Dyna-Soar.

"Man-in-space" was literally true. At the time, virtually no one in the United States or Soviet governments was thinking of sending women into space. The Soviets began training women for a spaceflight in 1962, and launched one of them the following year. But the second Soviet women didn't fly into space until 1982. U.S. women astronauts began training in 1978, and the first U.S. woman flew in 1983.

A few days after NASA officially came into being on October 1, 1958, its new administrator, T. Keith Glennan, gave the order that began the new man-in-space program. Choosing the people to run the new program was not difficult because a group of NACA engineers at Langley had been working almost full time for months on the problems of putting people in space. The Pilotless Aircraft Research Division (PARD) at Langley had worked for years on small rockets, developing in the process expertise in the problems of high-speed flight. Under the direction of a brilliant aeronautical engineer from Minnesota named Robert R. Gilruth, who had trained under balloonist Jean Piccard, PARD agreed to provide research expertise to the air force early in 1958 as the latter drew up plans to send humans into space. Many of the engineers at PARD, notably the head of its Performance Aerodynamics Branch, Maxime A. Faget, felt that PARD should have a bigger role in the new space program. The son of an army physician who was born in British Honduras, where his father was posted, Faget had earned an engineering degree at Louisiana State University and served during the war in the navy's submarine service before joining NACA. In early 1958, Faget advocated

for replacing winged vehicles or capsules that had some lifting characteristics with blunt-bodied ballistic capsules. As he outlined in a paper at a
conference at Ames in March, Faget believed that ballistic capsules simpli-
fied control and stabilization requirements in space and could be brought
home simply by firing retrorockets into the path of travel. Moreover, ballistic capsules could be made with greater ease and speed than winged vehicles like the X-20.

On November 5, 1958, NASA formally established the Space Task Group
(STG) at Langley to run the new man-in-space program. It put Gilruth in
charge of STG and the new program, which in December was named Project
Mercury, after the messenger of Greek mythology. The name was chosen
over another proposal, Project Astronaut, which was felt to put too much
focus on the men in the spacecraft. The initial nucleus of STG, which would
eventually grow into today's Johnson Space Center in Houston, Texas, was
formed from engineers and other personnel from PARD, from elsewhere in
Langley, and from the Lewis Research Center. Under Gilruth's direction,
Faget set to work designing a blunt capsule for Mercury, while others grappled with issues such as spacesuits, systems for the capsule, boosters, and
spacecraft tracking.

With competition from the Soviet Union looming, the leaders of NASA
and Project Mercury decided that the Redstone rocket would boost Mercury spacecraft in a series of suborbital flights, followed by orbital flights
launched by Atlas rockets. A small rocket built from solid-fuel rockets called
Little Joe was developed for early tests of the capsule and the launch escape
system. STG conducted a competition for the business of building the Mercury spacecraft, awarding the contract in January 1959 to McDonnell Aircraft Corporation of St. Louis, Missouri.

A crucial issue for Project Mercury was deciding who would fly or ride in
the Mercury spacecraft. A NASA committee drafted job descriptions in December 1958, but Gilruth and his fellow managers soon realized they wanted
military test pilots rather than civilian experts. Robert Voas, a psycholo-
gist who helped make the decision, said, "the selection of trained test pilots
would assure that the individuals would have a high level of stress tolerance and ability to perform effectively in emergencies." Gilruth agreed and
noted that test pilots already had substantial technical knowledge. "They
are used to altitude, the need for oxygen, bends and acceleration," he added.

"They are used to discipline and to taking risks, so I always felt that we should draw from professional aviators." Glennan and his deputy administrator, Hugh Dryden, also agreed, and saw additional benefit from the fact that military pilots had already passed their security clearances. Glennan and Dryden took their decision to President Eisenhower, who endorsed it "readily and without equivocation." The candidates were known variously as pilots or engineer-pilots, and more whimsically as spacemen or space pilots. The NASA committee drawing up the job description settled on the word *astronaut*, based on the already known terms *aeronaut* and *astronautics*. It turned out that the word *astronaut*, which means star sailor, had already appeared in science fiction as early as 1929.

Mercury's astronaut selection process began in January 1959, and candidates drawn from the ranks of air force, navy, and marine test pilots were asked to volunteer if they met requirements that included an age under forty; height under 5 feet, 11 inches; a bachelor's degree or the equivalent; matriculation at a test pilot school; a minimum total of fifteen hundred hours' flying time; jet pilot qualifications; and excellent physical condition. More than one hundred men met these requirements, and those who volunteered for selection underwent interviews and a battery of medical and psychiatric tests to reduce the number of candidates to a handful. The medical tests, which took place at the Lovelace Clinic, became legendary for their thoroughness. "They found openings in my body I didn't know I had," one participant later said. On April 9, 1959, NASA publicly announced the identities of seven men who would be known as astronauts: Lt. Cmdr. Alan B. Shepard, Lt. Malcolm Scott Carpenter, and Lt. Cmdr. Walter M. Schirra of the navy; Capt. Virgil I. "Gus" Grissom, Capt. Donald K. "Deke" Slayton, and Capt. Leroy Gordon Cooper of the air force; and Lt. Col. John H. Glenn of the marines. Perhaps the greatest ordeal for the "Mercury 7" was the Washington press conference where they were introduced. Flashbulbs popped throughout the event, and astronauts were quizzed about their families, their fears, and their religion. Then the press applauded them. "By the next morning, the seven Mercury astronauts were national heroes," Tom Wolfe wrote in his book *The Right Stuff.* "Even though so far they had done nothing more than show up for a press conference, they were known as the bravest men in America." The seven pilots were celebrities, and they had to spend time managing that celebrity. Within moments of the press conference, reporters were pursuing their families and acquaintances for details about these new heroes. Though accounts of their flights were open to any media organization, the astronauts sold their personal stories to *Life* magazine so as to manage the media intrusions into their lives.

With many aircraft and missile programs in full swing in 1959, NASA's wish to protect its existing research programs, and the perception that spaceflight was just a flash-in-the-pan experiment, STG had a difficult time attracting new engineers to the Mercury program. Fortunately for STG, the Canadian government canceled its plans to build a new jet interceptor called the CF-105 Avro Arrow early in 1959. About two thousand engineers were put out of work, and STG was able to hire thirty-one of Avro Canada's top engineers. These experts as well as newly graduated engineers from around the United States helped fill out STG's management and computer staff, and played a major role in establishing the control center and worldwide tracking network needed to follow the flights of Mercury and the spacecraft that flew afterward. The engineers of Project Mercury, whose experience and training had centered on building and flying aircraft, were setting out on a new field of endeavor: designing the first spacecraft and the infrastructure needed to support them.

NASA's decision to select test pilots to fly Mercury had an impact on the spacecraft design. Once the new astronauts began to learn about their spacecraft, they grew concerned about the lack of control they would have over it. Some people in NASA believed that the astronauts should simply be passengers in an automated spacecraft until it was proved that humans could withstand the rigors of flight. The test pilots hired as astronauts insisted on a spacecraft that they could control, and they got it. Early Mercury designs did not have a window that permitted the pilot to see where he was going. Nor did they have a hatch that the pilot could open when he chose. Both were incorporated in Mercury when the astronauts insisted on them. That fall, Slayton delivered a paper at the annual meeting of the Society of Experimental Test Pilots in which he asserted that flying Mercury required a test pilot on board, not just anyone from a "college-trained chimpanzee to the village idiot." Slayton's statements were a direct answer to the jibes of pilots like Chuck Yeager that astronauts "would have to sweep the monkey shit off the seat" before flying into space, a reference to the early primate launches.

A similar decision had to be made when the Project Mercury engineers pondered how to track and help control the spacecraft from the ground. Fred Matthews, who worked with Walter Williams and Christopher Kraft when they set up Mercury's ground systems, remembered that one contractor suggested an automated system to track and control Mercury. Kraft reacted instantly to the proposal, Matthews remembered, by explaining how he saw it working: "He would have a desk and on it there would be a phone, a green light and a red light. The green light would come on when everything was working the way it was supposed to work, and that would soothe his ulcers. If things weren't working the way they were supposed to work, the red light would come on saying all is lost, he might as well pick up the phone and call his wife and tell her he'll be home for lunch. That was the last we ever heard of that proposal." Williams and Kraft settled on assembling the team that formed the Mercury Control Center in Cape Canaveral, the forerunner of Mission Control in Houston, and the worldwide tracking network that kept the controllers in touch with the spacecraft.

Soviet Efforts

The Soviet Union's space experts also considered building winged spacecraft for their human passengers, but they never went as far in that direction as the United States. In Germany at war's end, the Soviets discovered a copy of a report written for the German military on a rocket-powered antipodal bomber by Eugen Sänger and his wife, Irene Bredt. Stalin commissioned engineers at the NII-I institute under the direction of mathematician Mstislav Keldysh to work on a Soviet antipodal bomber. Stalin's interest in the idea was so strong that the Germans who had been brought to Russia were asked for help, and Soviet agents even tried to kidnap Sänger from his postwar home in France. The kidnapping attempt failed when the officer sent to carry out the task defected to the West. The effort to develop the bomber eventually foundered because of the great leaps in engine and heat-protection technology that were required.

The news in 1957 that the U.S. Air Force was working on Dyna-Soar spurred three Soviet design bureaus to start work on rocket planes. A group led by Pavel Tsybin designed a space plane that Sergei Korolev nicknamed the Sandal because of its shape, but a combination of problems with the craft's heat-protection system and political interference led to its cancellation in 1959. Vladimir Myasishchev's M-48 spaceplane project collided with Vladimir Chelomei's ambitions to have his design bureau build a spaceplane. Chelomei had started building cruise missiles for the air force and navy after the war, but these lost favor to ICBMs. Chelomei then decided to pursue ambitious plans for space vehicles. Although he had several enemies in powerful positions, especially Dmitry Ustinov, he prospered during the Khrushchev era through his connections to the Soviet premier, including Khrushchev's son Sergei, an engineer who worked inside Chelomei's design bureau. Chelomei proposed a rocket plane of his own, called Raketoplan, and this project went ahead, only to be canceled in 1965 after the termination of Dyna-Soar and the fall of Khrushchev.

Sergei Korolev had long been thinking about sending humans into space, as his encouragement of Mikhail Tikhonravov's work in the late 1940s had shown. To test the idea that living beings could survive rocket flights, Korolev ordered nine more dog flights on modified R-1 rockets from 1954 through 1956, which showed that the dogs' breathing and physical functions weren't seriously affected by the rigors of flight. The dogs wore spacesuits and flew in special capsules equipped with cameras that followed their reactions to flight. Unfortunately, some of the dogs and other animals such as rabbits and rats that flew on these flights died when equipment failed. Korolev didn't conceal his desires when he spoke in public, as he did in September 1955: "Our mission is to ensure that a Soviet man becomes the first to fly in a rocket. And our mission is to ensure that it is Soviet rockets and Soviet spaceships that are the first to master the limitless space of the cosmos."

Korolev spoke those words with the knowledge that engineers and designers in his bureau were already drawing up suborbital spacecraft launched on uprated R-2 rockets that could carry one passenger. Capsules were being designed to return to Earth by various means, including parachutes, wings, braking engines, and even helicopter rotors that would deploy on the way down. Between 1957 and 1960, the R-2 carried more dogs to high altitudes in seven successful flights. One of the containers used in these flights was recycled for use in *Sputnik 2* to carry Laika on her flight into orbit that fall. By then, Korolev was talking about sending men into orbit and dispensing with suborbital flights altogether. On February 15, 1958, he ordered Tikhonravov to begin work on a piloted spacecraft for orbital flight, which was code-named Object K. Tikhonravov gave the engineering work to a group led by thirty-two-year-old engineer Konstantin Feoktistov, who had been shot by German troops and left for dead during World War II while serving as a scout for Soviet partisans.

Early in 1959, with the knowledge that the Americans had established NASA and Project Mercury, the Soviet government and its ruling Communist Party authorized work on piloted spacecraft and reconnaissance satellites, a matter of priority to the military. Korolev had separate groups working on both programs, but his resources were limited, so he decided to combine the efforts because the first Soviet reconnaissance satellites had to return to Earth with their precious photos much as did piloted spacecraft and the CORONA satellites being built in the United States. For that reason, Vostok, the first Soviet piloted spacecraft, shared many features with Zenit, the first Soviet reconnaissance satellite. Using computations that recommended ballistic reentries, Feoktistov and his team began by designing a spherical return capsule for their vehicles. The sphere was covered with a plastic material reinforced with asbestos fabric as a heat shield. Korolev wanted new recovery methods, but the capsule was soon fitted with traditional parachutes for recovery. Because the capsule was too heavy to land softly with the parachutes, it was also equipped with an ejection seat that would carry the passenger away to land under his or her own parachute before impact. The instrumentation that was needed during launch and in space, including a retrorocket, was carried outside the spherical capsule in a pressurized instrument module.

In early 1959, the Soviet air force and Korolev's design bureau began preparing to select passengers for the new spacecraft. Though their criteria for what they would call cosmonauts did have features in common with the American selection process, important differences reflected the differences between societies and space programs. The Soviet selection process did not begin until the identities of the Mercury astronauts were known, and the Soviets consciously decided that their trainees for space would be pilots, but not necessarily test pilots, and aged between twenty-five and thirty years, younger than their American counterparts. Korolev said he didn't need test pilots with engineering degrees like the Mercury astronauts because "our technology is such that we do not require, as the American Mercury Project does, that our early cosmonauts be highly skilled engineers." Because

29. The original 1960 group of cosmonauts is shown in a photo from May 1961 at the seaside port of Sochi. The names of many of these men were considered state secrets for more than twenty-five years. Sitting in front from left to right: Pavel Popovich, Viktor Gorbatko, Yevgeni Khrunov, Yuri Gagarin, Chief Designer Sergei Korolev, his wife Nina Koroleva with Popovich's daughter Natasha, Cosmonaut Training Center director Yevgeni Karpov, parachute trainer Nikolai Nikitin, and physician Yevgeni Fedorov. Standing the second row from left to right: Alexei Leonov, Andrian Nikolayev, Mars Rafikov, Dmitri Zaikin, Boris Volynov, Gherman Titov, Grigori Nelyubov, Valeri Bykovsky, and Georgi Shonin. In the back from left to right: Valentin Filatev, Ivan Anikeyev, and Pavel Belyayev. Four cosmonauts were missing from the photograph. Anatoli Kartashov and Valentin Varlamov and had both been dropped from training because of injuries. Valentin Bondarenko died in a training accident a few months before. Vladimir Komarov was indisposed. (NASA photo)

the new spacecraft was doubling as a reconnaissance satellite, it could operate without inputs from pilots.

About two hundred pilots met the criteria for cosmonaut selection, and these men were brought to Moscow for rigorous medical exams and physical tests that included rides in centrifuges and exposure to different air pressures. Late in 1959, the list of candidates was winnowed down to twenty men. Air force lieutenant general Nikolai Kamanin, who was a decorated hero for his daring airborne rescue of crew members from a ship trapped in arctic ice in 1934, was chosen to direct the team of cosmonauts. In February 1960, he ratified the selections of twenty junior air force officers to train for the team: Ivan Anikeyev, Pavel Belyayev, Valentin Bondarenko, Valeri Bykovsky, Valentin Filatev, Yuri Gagarin, Viktor Gorbatko, Anatoli Kartashov, Yevgeni Khrunov, Vladimir Komarov, Alexei Leonov, Grigori Nelyubov, Andrian

Nikolayev, Pavel Popovich, Mars Rafikov, Georgi Shonin, Gherman Titov, Valentin Varlamov, Boris Volynov, and Dmitri Zaikin.

The group reported for training near Moscow and began their preparations for space flight, which stressed parachute training. After an initial round of training and the arrival of a basic spacecraft simulator in May, the air force reduced the size of the training group for the first flight to six: Gagarin, Kartashov, Nikolayev, Popovich, Titov, and Varlamov. Kartashov and Varlamov were dropped for medical reasons that summer and were replaced by Bykovsky and Nelyubov in the group of six candidates preparing for the first flight.

A big difference between the American and Soviet training process was the anonymity of the Soviet trainees prior to their flights. Their names were kept secret until after they had entered space, and the identities of many of the first group of cosmonaut trainees remained unknown until the collapse of the Soviet Union. Similarly, though the first Soviet cosmonaut trainees were trained in controlling their spacecraft, they were to touch the controls only with the consent of the ground or in an extreme emergency. This fact was also kept under wraps for years after the first flights.

Test Launches

By the time the prospective astronauts and cosmonauts began training, both the U.S. Mercury and Soviet spacecraft were undergoing flight testing. In 1959 and 1960, Mercury capsules were test-flown atop Little Joe boosters at Wallops Island, Virginia, and on Redstone and Atlas rockets at Cape Canaveral. Two rhesus monkeys, Sam and Miss Sam, flew on Little Joes while the Mercury launch escape system underwent testing, and both were safely recovered. Though many launches succeeded, a Mercury capsule was lost in 1960 when its Atlas booster disintegrated early in flight. An attempt to launch a Mercury capsule atop a Redstone ended in embarrassment when the rocket rose two inches, shut down, and then settled back into place while the escape tower flew away and released the capsule's parachute. The fully armed rocket stood alone on the launch pad for hours until its batteries ran out. A suggestion that the rocket be depressurized with a shot to its tanks was rejected, in contrast to a similar situation years before when a rifle shot depressurization had saved a Viking rocket at White Sands.

The Mercury team soon succeeded with another Redstone, however, and the second Mercury-Redstone flight was cleared to carry the male chimpanzee Ham in a dress rehearsal for the first Mercury flight with a man. On January 21, 1961, Ham was taken out to the launch pad at Cape Canaveral and strapped into his custom-made couch inside the Mercury capsule. Ham's arms were left free so he could perform his assigned tasks. Three lights and three levers were located just above his couch, and metal plates were attached to the underside of his feet just as they had been during training. He had to push the lever before one of the lights went out or receive a shock. After several delays, the *Mercury-Redstone 2* flight lifted off, and Ham began pushing his levers as he had been trained to do. Unfortunately, a thrust regulator on the Redstone jammed open, and the rocket traveled at a greater speed than expected, seriously throwing off the splashdown calculations. Eight navy ships were waiting in the Atlantic recovery area 290 miles out to sea, but the capsule, having reached a higher-than-planned altitude of 156.5 miles, overshot its target by nearly 120 miles. The capsule parachuted down and landed in the Caribbean Sea near the Bahamas after an eighteen-minute flight. Despite pulling an unexpected load of around 18 g's during the launch, Ham had performed exceptionally well. A recovery ship finally reached the bobbing capsule three hours after splashdown and winched it out of the sea. When Ham was removed from his capsule, he eagerly devoured an apple as a postflight snack. He was in good shape and went on to live a long life.

Ham's flight took place during a nervous time for NASA. The day before, John F. Kennedy, a Democrat, had replaced the Republican Eisenhower as president, and Kennedy's attitudes toward NASA and space were the subject of some speculation. During the 1960 election campaign, Kennedy's campaign rhetoric criticized Eisenhower's space policy for not being aggressive enough. But after the election, Kennedy's new science advisor, Jerome B. Wiesner of MIT, had written and released a report that was critical of NASA in general and Project Mercury in particular. "We should stop advertising Mercury as our major objective in space activities," Wiesner's report said, echoing a strong current of criticism from prominent scientists. "We should find effective means to make people appreciate the cultural, public service, and military importance of space activities other than space travel." NASA administrator Glennan, also a Republican, had stepped down when Eisen-

hower's term ended, and his replacement, James E. Webb, a former director of the Bureau of the Budget under President Harry Truman, took over
in February. Many wondered whether Webb, who had virtually no background in aviation or science, would make a good leader for NASA.

NASA began studies in 1959 on piloted space programs to follow Mercury,
identifying flight to the moon as a logical target for the next human spaceflight program. In 1960, NASA announced that a program called Apollo
would follow Mercury. This spacecraft would fly atop the Saturn rockets
being built by von Braun's group in Huntsville, and Apollo's tentative goal
was a flight around the moon. Many people within NASA, including von
Braun, were examining the technical issues involved in landing astronauts
on the moon, but Eisenhower withheld budgetary approval for these ambitious plans in his final budget proposal, sent to Congress just days before
he left office. In his farewell address to the nation, Eisenhower did not discuss space, but famously warned against the growing influence of the "military-industrial complex" and the "scientific-technological elite."

A month after Ham's flight, a Mercury capsule flew successfully atop an
Atlas booster, much to the relief of Project Mercury management who had
seen the previous Mercury-Atlas disintegrate shortly after launch. Although
the next launch on their schedule was a Mercury-Redstone launch with an
astronaut on board, von Braun and his staff were greatly concerned by the
thrust regulator problem on the Redstone that caused MR-2 to overshoot
its target. Moreover, Ham's capsule had sustained serious damage while it
bounced around in the rough seas awaiting recovery and would have to be
repaired before astronauts flew. Though many within NASA, notably the astronauts, were anxious to start flying, von Braun and one of his top lieu-
tenants, Kurt Debus, felt strongly that another Mercury-Redstone should
fly before an astronaut stepped on board. In February 1961, Gilruth and
NASA officials in Washington made the fateful decision to allow the additional Redstone flight and postpone the first flight with an astronaut until late April. On March 24, the mission known as MR-BD, for Booster Development, lifted off the launch pad at Cape Canaveral and demonstrated
that the Redstone's thrust regulator problem had been fixed.

The celebrations of this success were overshadowed the next day by news
that the Soviet Union had launched and recovered a spacecraft that was
clearly a test for its human space program. *Korabl-Sputnik 5*, which flew

on March 25, was just the latest in a series of test flights for the new Soviet spacecraft that would be called Vostok (East). The spacecraft was developed under great pressure. Some of it was coming from Korolev's determination to get a Soviet cosmonaut in orbit before the first manned Mercury-Redstone suborbital flight. Other pressure emanated from the Soviet government, which had watched the U.S. Discoverer test flights for the CORONA reconnaissance satellites with growing concern and wanted Soviet Zenit reconnaissance vehicles operating as soon as possible. The first prototype of Vostok and Zenit was launched into orbit on May 15, 1960, and was called *Korabl-Sputnik 1*, or "satellite-ship 1." This first spacecraft was not designed to return intact to Earth, but a retrorocket test that was supposed to burn up the vehicle in the atmosphere instead sent it into a higher orbit. The next launch on July 28 carried two dogs and a recovery system, but a rocket failure shortly after launch doomed the canines, Chayka and Lischka. *Korabl-Sputnik 2* carried the dogs Strelka and Belka into and back from space after a day in orbit on August 19, making the two dogs the first living beings brought home from orbit. The next flight in the series, on December 1, entered orbit successfully, but the retrorocket firing the next day partially failed. The dogs Pchelka and Mushka died during *Korabl-Sputnik 3*'s delayed reentry when a self-destruct system designed to prevent the spacecraft from falling into foreign hands destroyed the spacecraft. The next launch attempt three weeks later turned into a suborbital abort when the third stage failed, but the dogs Kometa and Shutka were spared when the self-destruct apparatus also malfunctioned. The dogs were trapped in the spacecraft for days, however, because it landed in deep snow in an isolated part of the Soviet Union. Just as the Americans were facing a string of booster problems with Mercury, the Soviets had to deal with two failed Vostok test flights in a row.

Korolev and his team worked on the problems afflicting their spacecraft and launched the first human-rated Vostok on March 9, 1961. *Korabl-Sputnik 4* carried one dog, Chernuska; an assortment of other animals, including mice, guinea pigs, and reptiles; and a variety of biological samples. In the craft's ejection seat sat a mannequin dressed in the orange Soviet pressure suit. After a one-orbit flight, the spacecraft safely returned to Earth. The feat was repeated on March 25, the day after the MR-BD flight. Although bad weather delayed the rescue of Zvezdochka, the single canine passenger of

Korabl-Sputnik 5, the flight was a success. The launch was witnessed by the six cosmonauts who were in training for the first manned flight. Although the way was now clear for one of them to fly into space, a shadow had been cast on the successful tests by the tragic death of cosmonaut trainee Valentin Bondarenko two days earlier on March 23. During an isolation test in a chamber with a high level of oxygen, Bondarenko had accidentally thrown a cotton swab onto a live electric hotplate. The swab burst into flames, and so did the oxygen-rich atmosphere of the isolation chamber. His body covered with burns, Bondarenko died eight hours later.

In spite of this tragedy, which was kept secret for years, preparations went on for the first flight into space with a cosmonaut on board. The Russians concealed their flight preparations, exploiting their knowledge that the Americans were planning their first Mercury-Redstone suborbital flight for no earlier than late April. Finally, on the morning of Wednesday, April 12, 1961, the man selected to make the flight, twenty-seven-year-old air force senior lieutenant Yuri Alexeyevich Gagarin, donned his orange spacesuit and white helmet, took a bus to the same launch pad where Sputnik had been launched more than three years earlier, embraced Korolev and other officials, and then boarded his spacecraft. At 9:07 a.m. Moscow time, the R-7 rocket roared away from the pad. In addition to the usual telemetry from the rocket and the spacecraft, the engineers in the control center heard a confident voice: "Poyekhali!" ("We're off!") Humankind had entered a new age of exploration.

Epilogue: July 16, 1969

In the roads, beaches, islands, and waters surrounding a space center in Florida named after John F. Kennedy, an estimated one million people had gathered to watch the sixth launch of Wernher von Braun's greatest creation, the Saturn V rocket. Every launch of this 363-foot behemoth was an event due to the 7.5 million pounds of thrust it packed at liftoff. But this particular flight was even more special because astronauts were on board who would take the *Apollo 11* spacecraft all the way to the surface of the moon for the first human landing on another world. Though this landmark flight began, in a sense, with humanity's earliest urgings to travel beyond the earth, it had its more immediate origins in the flight of Yuri Gagarin.

The Soviet cosmonaut's successful 108-minute flight on April 12, 1961, was the opening event of an extraordinary six weeks in spaceflight history that led directly to *Apollo 11* astronauts Neil Armstrong and Buzz Aldrin stepping on the moon eight years, three months, and eight days later. Gagarin's single orbit around the earth was followed on May 5 by Alan Shepard's 15-minute suborbital flight aboard his Mercury-Redstone vehicle. President Kennedy, whose young administration was shaken by Gagarin's flight and a botched invasion of Fidel Castro's Cuba a few days later, decided to make a bold announcement to show American determination to be first in space. On May 25, the president set a national goal of "landing a man on the moon and returning him safely to the earth" by the end of the 1960s.

Led by men like von Braun, Robert Gilruth, James Webb, and George Mueller, NASA successfully pulled off six manned Mercury and ten Gemini flights. After overcoming the setback of a 1967 launch pad fire that killed the first Apollo crew—astronauts Virgil Grissom, Edward White, and Roger Chaffee—NASA succeeded in flying four manned Apollo flights before fulfilling President Kennedy's national goal. In July 1969 Armstrong and Aldrin reached the lunar surface and returned home safely with astronaut Michael Collins. Five more Apollo crews journeyed to the lunar surface and another three crews flew to the vicinity of the moon before Apollo ended with *Apollo 17* in December 1972.

When *Apollo 11* departed skyward from the Kennedy Space Center, many of the people who had made the flight possible were among the crowds nearby. Many of NASA's top officials were in the firing room of Launch Complex 39, including von Braun. At a press briefing before the launch, he was asked how he rated the importance of *Apollo 11*. "I think it is equal in importance to that moment in evolution when aquatic life came crawling up on the land," he said.

At the time of *Apollo 11*, von Braun was drawing up ambitious plans for NASA to build a shuttle craft, a space station, a lunar base, and spacecraft to go to Mars in the 1980s. His plans were boosted when Vice President Spiro Agnew returned from watching the *Apollo 11* launch to tell assembled media that America should go to Mars. President Richard Nixon, who began his term a few months earlier in 1969, rejected von Braun's ambitious plans but eventually agreed to a shuttle. A frustrated von Braun left NASA for private industry as Apollo wound down in 1972 and died five years later, advocating space exploration to the end.

Many of the rocketeers who came with von Braun to America from Peenemünde were also there at the *Apollo 11* launch site or standing by at the Marshall Space Center in Huntsville. A number of them held highly responsible positions, such as Kurt Debus, who was director of the space center from which *Apollo 11* was launched, and Eberhard Rees, who succeeded von Braun as director of Marshall a few months after the historic lunar flight.

Arthur C. Clarke, the great British writer, theorist, and spaceflight advocate, also came to Florida and told media covering the launch: "This is the last day of the old world." Willy Ley, who was one of the key publicizers of space travel in his native Germany and then his adopted United States, died eight weeks before the launch he had done so much to make a reality.

Also among those at the space center was Hermann Oberth, the only one of the three fathers of space exploration still living. Oberth had lived in the United States for three years in the late 1950s as an advisor to von Braun at the Army Ballistic Missile Agency, but he returned to Germany to retire, and died in 1989 at the age of ninety-five.

Konstantin Tsiolkovsky had died in 1935, and none of the Russians who followed in his footsteps were in Florida that day because the Cold War

30. Dr. von Braun and Prof. Hermann Oberth are honored by the Berlin Technical University. Both received honorary doctorates on January 8, 1963. (NASA photo)

competition both restricted travel and kept their own launch site off limits to almost all foreigners. Indeed, when Gagarin was launched from Tyura Tam, the Soviets identified his launch site as Baikonur, the name of a distant village, in an unsuccessful attempt to deceive their adversaries. Gagarin was killed in 1968 in a flying accident, and Sergei Korolev had died in 1966 on the operating table. It was only after Korolev's death, when he was given a hero's funeral in Red Square, that his role as chief designer of the Soviet space program was finally acknowledged in public. At the time of his death, Korolev was organizing the Soviet effort to put a man on the moon before Apollo. The effort fell short because of his death and the low levels of government support that forced Korolev's gigantic N-1 rocket to be launched without proper testing. The N-1 failed four times before the Soviet moon program was canceled. The USSR itself collapsed in 1991, and soon afterward the competition that marked the Cold War years gave way to cooperation that saw Russians fly from the same launch pad as *Apollo 11,* and Americans rocket from the launch site of Sputnik and Gagarin.

Robert Goddard had been dead for nearly twenty-four years in 1969, but his friend and benefactor Charles Lindbergh was present at the launch site. Goddard's wife, Esther, told a reporter from her home in Worcester, Massachusetts, that Goddard "would have been beside himself with delight" at the launch of *Apollo 11*. His dream, which was the dream of so many others, was coming true that week. The years to come would make other dreams a reality as robot spacecraft visited and in some cases touched other bodies throughout the solar system. But other space dreams, such as the permanent occupation of the moon and human flights to Mars, would have to wait.

Sources

Atkinson, Joseph D., Jr. and Jay M. Shafritz. *The Real Stuff: A History of* NASA's *Astronaut Recruitment Program*. New York: Praeger, 1985.

Bainbridge, William Sims. *The Spaceflight Revolution: A Sociological Study*. Reprint edition. Malabar FL: Robert E. Krieger Publishing, 1983.

Baker, David. *The Rocket: The History and Development of Rocket and Missile Technology*. New York: Crown Publications, 1978.

———. *Spaceflight and Rocketry: A Chronology*. New York: Facts on File, 1996.

Béon, Yves. *Planet Dora: A Memoir of the Holocaust and the Birth of the Space Age*. Boulder CO: Westview Press, 1997.

Bergaust, Erik. *Wernher von Braun*. Washington DC: National Space Institute, 1976.

Bergonzi, Bernard, ed. *H. G. Wells: A Collection of Critical Essays*. Englewood Cliffs NJ: Prentice Hall, 1976.

Bille, Matt, and Erika Lishock. *The First Space Race: Launching the World's First Satellites*. College Station: Texas A&M University Press, 2004.

Bilstein, Roger E. *Stages to Saturn: A Technological History of the Apollo/Saturn Launch Vehicle*. Washington DC: National Aeronautics and Space Administration, 1980.

Brooks, Courtney G., James M. Grimwood, and Loyd S. Swenson. *Chariots for Apollo: A History of Manned Lunar Spacecraft*. Washington DC: National Aeronautics and Space Administration, 1979.

Buckbee, Ed, and Wally Schirra. *The Real Space Cowboys*. Burlington ON: Apogee Books, 2005.

Burgess, Colin. "The Prelude: Sputniks, Space Dogs and Astrochimps." Unpublished draft, 2003.

Burrows, William E. *Deep Black: Space Espionage and National Security.* New York: Berkeley Books, 1988.

―――. *This New Ocean: the Story of the First Space Age.* New York: Random House, 1998.

Caro, Robert A. *The Years of Lyndon Johnson: Master of the Senate.* New York: Alfred A. Knopf, 2002.

Carpenter, Scott, and Kris Stoever. *For Spacious Skies: The Uncommon Journey of a Mercury Astronaut.* Orlando FL: Harcourt, 2002.

Chang, Iris. *Thread of the Silkworm.* New York: Basic Books, 1995.

Chertok, Boris. *Rockets and People.* Vol. 1. Washington DC: National Aeronautics and Space Administration, 2005.

Clary, David A. *Rocket Man: Robert H. Goddard and the Birth of the Space Age.* New York: Hyperion, 2003.

Daniloff, Nicholas. *The Kremlin and the Cosmos.* New York: Alfred A. Knopf, 1972.

DeChambeau, Aimée. "Struggles of the Father." *Ad Astra* (September/October 2002): 41–44.

Dickson, Paul. *Sputnik: The Shock of the Century.* New York: Walker and Company, 2001.

Divine, Robert A. *The Sputnik Challenge: Eisenhower's Response to the Soviet Satellite.* Oxford: Oxford University Press, 1993.

Dornberger, Walter. *V-2.* New York: Bantam Books, 1979.

Durant, Frederick C., and George S. James, eds. *First Steps toward Space: First and Second* IAA *History Symposia.* San Diego: American Astronautical Society, 1985.

Dyson, George. *Project Orion: The True Story of the Atomic Spaceship.* New York: Henry Holt, 2002.

"Edgar Rice Burroughs 1875–1950." http://www.kirjasto.sci.fi/erburrou.htm. Accessed March 15, 2004.

Emme, Eugene M., ed. *The History of Rocket Technology.* Detroit: Wayne State University Press, 1964.

Freedman, Russell. *Jules Verne: Portrait of a Prophet.* New York: Holiday House, 1965.

Freeman, Marsha. *How We Got to the Moon: The Story of the German Space Pioneers.* Washington DC: 21st Century Science Associates, 1993.

Gainor, Chris. *Arrows to the Moon: Avro's Engineers and the Space Race.* Burlington ON: Apogee Books, 2001.

————. "The Chapman Report and the Development of Canada's Space Program." *Quest: The History of Spaceflight Quarterly* 10, no. 4 (2003): 3–19.

Garlinski, Jozef. *Hitler's Last Weapons: The Underground War against the V1 and V2.* New York: Times Books, 1978.

Gernet, Jacques. *A History of Chinese Civilization.* Cambridge: Cambridge University Press, 1982.

Glushko, Alexander. "Ivan T. Kleimyonov: A Talented Organizer." *Quest: The History of Spaceflight Quarterly* 8, no. 3 (2000): 24–31.

Goddard, Esther C., and G. Edward Pendray, eds. *The Papers of Robert H. Goddard.* 3 vol. New York: McGraw-Hill, 1970.

Godwin, Robert, ed. *Dyna-Soar: Hypersonic Strategic Weapons System.* Burlington ON: Apogee Books, 2003.

Goldsmith, Donald, and Tobias Owen. *The Search for Life in the Universe.* 3d ed. Sausalito CA: University Science Books, 2001.

Golovanov, Yaroslav. *Russians in Space—40th Anniversary of Sputnik.* CD-ROM. Los Angeles/Moscow: Ultimax, 1997.

Gray, Mike. *Angle of Attack: Harrison Storms and the Race to the Moon.* New York: W. W. Norton, 1992.

"The Great Moon Hoax." http://www.museumofhoaxes.com/moonhoax .html. Accessed March 14, 2004.

Green, Constance M., and Milton Lomask. *Vanguard—A History.* Washington DC: National Aeronautics and Space Administration, 1970.

Gruntman, Mike. *Blazing the Trail: The Early History of Spacecraft and Rocketry.* Reston VA: American Institute of Aeronautics and Astronautics, 2004.

Hale, Edward Everett. *The Brick Moon.* http://www.readbookonline.net/ readOnLine/2108/. Accessed, March 14, 2004.

Hansen, James R. *Spaceflight Revolution: NASA Langley Research Center From Sputnik to Apollo.* Washington DC: National Aeronautics and Space Administration, 1995.

Harford, James. *Korolev: How One Man Masterminded the Soviet Drive to Beat America to the Moon.* New York: John Wiley & Sons, 1997.

Harmon, A. M., trans. *Lucian.* Vols. 1 and 2. Cambridge MA: Harvard University Press, 1979.

Harvey, Brian. *China's Space Program: From Conception to Manned Space-flight.* Chichester UK: Praxis Publishing, 2004.

———. "Yuri Kondratyuk: Stolen Past, Stolen Future?" *Quest: The History of Spaceflight Quarterly* 6, no. 4 (winter 1998): 31.

Heppenheimer, T. A. *A Brief History of Flight: From Balloons to Mach 3 and Beyond.* New York: John Wiley, 2001.

———. *Countdown: A History of Space Flight.* New York: John Wiley and Sons, 1997.

Hornblower, Simon, and Antony Spawforth, eds. *The Oxford Companion to Classical Civilization.* Oxford: Oxford University Press, 1998.

Hoskin, Michael, ed. *The Cambridge Concise History of Astronomy.* Cambridge: Cambridge University Press, 1999.

Jenkins, Dennis R. *Space Shuttle: The History of the National Space Transportation System—The First 100 Missions.* Cape Canaveral FL: Dennis R. Jenkins, 2001.

Koppes, Clayton R. *JPL and the American Space Program: A History of the Jet Propulsion Laboratory.* New Haven: Yale University Press, 1982.

Kosmodemyansky, A. *Konstantin Tsiolkovsky: His Life and Work.* Moscow: Foreign Languages Press, 1956.

Krige, John, and Arturo Russo. *Europe in Space 1960–1973.* Noordwjik, The Netherlands: ESA Publications Division, 1994.

Launius, Roger D. *NASA: A History of the U.S. Civil Space Program.* Malabar FL: Krieger Publishing, 1994.

———. *Space Stations: Base Camps to the Stars.* Washington DC: Smithsonian Books, 2003.

———, and Howard E. McCurdy, eds. *Spaceflight and the Myth of Presidential Leadership.* Urbana: University of Illinois Press, 1997.

Lehman, Milton. *This High Man: The Life of Robert H. Goddard.* New York: Farrar, Strauss, 1963.

Le Maner, Yves, and André Sellier. *Images de Dora 1943–1945: Voyage au Coeur du IIIe Reich.* St. Omer, France: La Coupole Editions, 1999.

Ley, Willy. *Rockets, Missiles, and Men in Space.* New York: Viking, 1968.

Logsdon, John M., ed. *Exploring the Unknown: Selected Documents in the History of the U.S. Civil Space Program.* Volume 1, *Organizing for Exploration.* Washington DC: National Aeronautics and Space Administration, 1995.

Lytkin, Vladimir, with Ben Finney and Liudmila Alepko. "Tsiolkovsky, Russian Cosmism and Extraterrestrial Intelligence." *Quarterly Journal of the Royal Astronomical Society* 36 (1995): 369–76.

Macinnis, Peter. *Rockets: Sulfur, Sputnik and Scramjets.* Crows Nest NSW Australia: Allen & Unwin, 2003.

Matthews, C. Frederick. Interview by author. Lexington, Massachusetts. October 12, 1997.

Maynard, Owen E. Interview by author. Tape recording and transcript. Waterloo, Ontario. May 9, 1995.

McCurdy, Howard E. *Space and the American Imagination.* Washington DC: Smithsonian Institution Press, 1997.

McDougall, Walter A. . . . *The Heavens and The Earth: A Political History of the Space Age.* New York: Basic Books, 1985.

Michel, Jean, with Louis Nucera. *Dora.* New York: Holt, Rinehart and Winston, 1979.

Miller, Jay. *The X-Planes: X-1 to X-45.* Hinckley, England: Midland Publishing, 2001.

Murray, Charles, and Catherine Bly Cox. *Apollo: The Race to the Moon.* New York: Simon and Schuster, 1989.

Needell, Allan A. *Science, Cold War and the American State: Lloyd V. Berkner and the Balance of Professional Ideals.* Amsterdam: Harwood Academic Publishers, 2000.

Neufeld, Michael J. *The Rocket and the Reich.* Cambridge MA: Harvard University Press, 1996.

Ordway, Frederick I., and Mitchell R. Sharpe. *The Rocket Team.* London: William Heinemann, 1979.

Pendray, G. Edward. "The Man Who Ushered in the Space Age." In *Rocket Development: Liquid Fuel Rocket Research 1929–1941,* edited by Robert H. Goddard, xiii–xxi. New York: Prentice-Hall, 1961.

Petrovich, G. V., editor-in-chief. *The Soviet Encyclopedia of Space Flight.* Moscow: Mir Publishers, 1969.

Piszkiewicz, Dennis. *Wernher von Braun: The Man Who Sold the Moon.* Westport CT: Praeger, 1998.

Powell, Joel. "The NOTSNIK Program: The Top Secret Air-Launched Satellite Attempts of 1958." *Quest: The History of Spaceflight Quarterly* 3, no. 1 (spring 1994): 58–61.

Reeves, Robert. *The Superpower Space Race: An Explosive Rivalry through the Solar System.* New York: Plenum Press, 1994.

Reichman, Jim. "Russian Space Pioneer Yuri V. Kondratyuk." *Orbit: The Journal of the Astro Space Stamp Society,* no. 46 (June 2000): 24–27.

Riabchikov, Evgeny. *Russians in Space.* Garden City NJ: Doubleday, 1971.

Roberts, J. M. *The Pelican History of the World.* Middlesex: Penguin Books, 1984.

Rosen, Milton W. *The Viking Rocket Story.* London: Faber and Faber, 1955.

Ryan, Craig. *The Pre-Astronauts: Manned Ballooning on the Threshold of Space.* Annapolis MD: Naval Institute Press, 1995.

Scala, Keith J. "A History of Air-Launched Space Vehicles." *Quest: The History of Spaceflight Quarterly* 3, no. 1 (spring 1994): 34–41.

Sereny, Gitta. *Albert Speer: His Battle with Truth.* New York: Vintage Books, 1996.

Sheehan, William. *The Planet Mars: A History of Observation and Discovery.* Tucson: University of Arizona Press, 1996.

Siddiqi, Asif A. *Challenge to Apollo: The Soviet Union and the Space Race, 1945–1974.* NASA SP-2000–4408. Washington DC: National Aeronautics and Space Administration, 2000.

———. *Deep Space Chronicle: A Chronology of Deep Space and Planetary Probes 1958–2000.* NASA Monographs in Aerospace History No. 24. Washington DC: National Aeronautics and Space Administration, 2002.

———. "The Rockets' Red Glare: Technology, Conflict, and Terror in the Soviet Union." *Technology and Culture* 44, no. 3 (July 2003): 470–501.

Slayton, Donald K., and Michael Cassutt. *Deke! U.S. Manned Space: From Mercury to the Shuttle.* New York: Forge, 1994.

Smolders, Peter. *Soviets in Space.* New York: Taplinger Publishing, 1974.

Spires, David N. *Beyond Horizons: A Half Century of Air Force Space Leadership.* Honolulu: University Press of the Pacific, 2002.

Stoiko, Michael. *Pioneers of Rocketry.* New York: Hawthorn Books, 1974.

Stuhlinger, Ernst, and Frederick I. Ordway. *Wernher von Braun, Crusader for Space: A Biographical Memoir*. Malabar FL: Krieger Publishing, 1994.

Swenson, Loyd S., James M. Grimwood, and Charles C. Alexander. *This New Ocean: A History of Project Mercury*. Washington DC: National Aeronautics and Space Administration, 1966.

Thompson, Milton O. *At the Edge of Space: The X-15 Flight Program*. Washington DC: Smithsonian Institution Press, 1992.

"Topics of the Times" (Robert Goddard). *New York Times*, January 13, 1920.

Verne, Jules. *From the Earth to the Moon*. Trans. Edward Roth. Mattituck NY: Aeonian Press, 1976.

———. *Trip Around the Moon*. Project Gutenberg. http://www.gutenberg .org. Accessed February 20, 2004.

Volkman, Ernest. *Science Goes to War: The Search for the Ultimate Weapon*. New York: Wiley, 2002.

Von Braun, Wernher, and Frederick I. Ordway III. *The History of Rocketry and Space Travel*, Rev. ed. New York: Thomas Y. Crowell, 1969.

Walker, Chuck, with Joel Powell. *Atlas: The Ultimate Weapon*. Burlington, ON: Apogee Books, 2005.

Walt Disney Treasures. *Tomorrow Land: Disney and Space and Beyond*. DVD. Burbank CA: Buena Vista Home Entertainment, 2003.

Walters, Helen B. *Hermann Oberth: Father of Space Travel*. New York: Macmillan, 1962.

Wells, H. G. *The First Men in the Moon*. Project Gutenberg. http://www .gutenberg.org. Accessed March 2, 2004.

———. *The War of the Worlds*. Project Gutenberg. http://www.gutenberg. org. Accessed March 1, 2004.

Williamson, Jack. *H. G. Wells: Critic of Progress*. Baltimore: Mirage Press, 1973.

Winick, Lester E. "Birth of the Russian Space Programme." *Spaceflight* (London) 20, no. 5 (May 1978): 162–73.

Winter, Frank H. *Prelude to the Space Age: The Rocket Societies: 1924–1940*. Washington DC: Smithsonian Institution Press, 1983.

———. "The Silent Revolution: How R. H. Goddard Helped Start the Space Age." Paper (IAA.6.15.1) presented at the 55th Congress of the International Astronautical Federation, Vancouver, BC, Canada, October 4–8, 2004.

———. "Sir William Congreve: A Bi-Centennial Memorial." *Spaceflight* (London) 14, no. 9 (September 1972): 333–34.

———. "William Hale—A Forgotten British Rocket Pioneer." *Spaceflight* (London) 15, no. 1 (January 1973): 31–33.

Wolfe, Tom. *The Right Stuff.* New York: Farrar, Strauss, Giroux, 1979.

Yeager, Gen. Chuck, and Leo Janos. *Yeager: An Autobiography.* New York: Bantam Books, 1986.

Zak, Anatoly. "Konstantin Tsiolkovsky Slept Here." *Air & Space Smithsonian*, August/September 2002, 62–69.

———. "The Rest of the Rocket Scientists." *Air & Space Smithsonian*, September 2003, 68–74.

Index

In the Outward Odyssey: A People's History of Spaceflight Series

Into That Silent Sea
Trailblazers of the Space Era, 1961–1965
Francis French and Colin Burgess
Foreword by Paul Haney

In the Shadow of the Moon
A Challenging Journey to Tranquility, 1965–1969
Francis French and Colin Burgess
Foreword by Walter Cunningham

To a Distant Day
The Rocket Pioneers
Chris Gainor
Foreword by Alfred Worden

Homesteading Space
The Skylab Story
David Hitt, Owen Garriott, and Joe Kerwin
Foreword by Homer Hickam

UNIVERSITY OF NEBRASKA PRESS

Also of Interest

Into That Silent Sea
Trailblazers of the Space Era, 1961–1965

By Francis French and Colin Burgess
With a foreword by Paul Haney

Through dozens of interviews and access to Russian and American official documents and family records, *Into That Silent Sea* captures the intimate stories of the men and women who made the space race their own and gave the era its compelling character.

ISBN: 978-0-8032-1146-9 (cloth)

In the Shadow of the Moon
A Challenging Journey to Tranquility, 1965–1969

By Francis French and Colin Burgess
With a foreword by Walter Cunningham

In the Shadow of the Moon tells the story of the most exciting and challenging years in spaceflight, with two superpowers engaged in a titanic struggle to land one of their own people on the moon. While describing awe-inspiring technical achievements, the authors go beyond the missions and the competition of the space race to focus on the people who made it all possible. Their book explores the inspirations, ambitions, personalities, and experiences of the select few whose driving ambition was to fly to the moon.

ISBN: 978-0-8032-1128-5 (cloth)

Fallen Astronauts
Heroes Who Died Reaching for the Moon

By Colin Burgess and Kate Doolan, with Bert Vis
Foreword by Captain Eugene A. Cernan U.S. Navy (Ret.),
Commander, Apollo 17

This book enriches the saga of mankind's greatest scientific undertaking, Project Apollo, and conveys the human cost of the space race—by telling the stories of those sixteen astronauts and cosmonauts who died reaching for the moon.

ISBN: 978-0-8032-6212-6 (paper)

Order online at www.nebraskapress.unl.edu or call 1-800-755-1105. When ordering mention the code BOFOX to receive a 20% discount.